THE HISTORY OF
CLOCKS
AND WATCHES

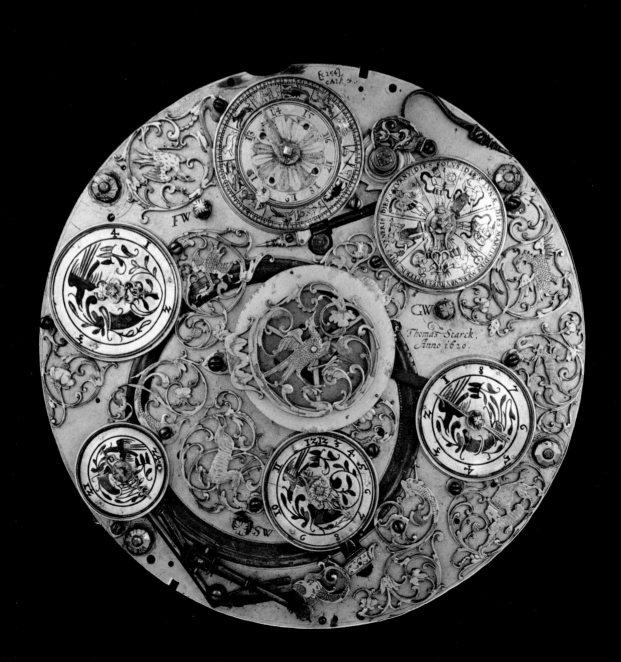

THE HISTORY OF
CLOCKS
AND WATCHES

Eric Bruton

RIZZOLI
NEW YORK

Published in the United States of America
in 1979 by

Rizzoli INTERNATIONAL PUBLICATIONS, INC.
712 Fifth Avenue/New York 10019

Library of Congress Catalog Card Number: 79-65958
ISBN: 0-8478-0261-2
Printed in Italy by IGDA, Novara

Half-title page : A watch of the last quarter of the seventeenth century with an experimental dial, known as a wandering hour. The hour numeral changes and its position indicates the minutes. Quarters are also shown under the semi-circular aperture. (By courtesy of the Trustees of the British Museum).

Title page : The dials of a fine monstrance clock in a case of inlaid tortoiseshell made in France c. 1680. The movement was made by Thomas Stark in 1620. The main dial, 240mm (9½ ins) in diameter has two double-XII dials on the front and several others on the back plate to provide various astronomical and astrological indications as well as calendar information. (By courtesy of the Trustees of the British Museum).

Contents page : A plate from a French treatise on horology by Claudius Saunier, first published in 1861, and translated into English. It shows part of the mechanism of a repeater watch.

CONTENTS

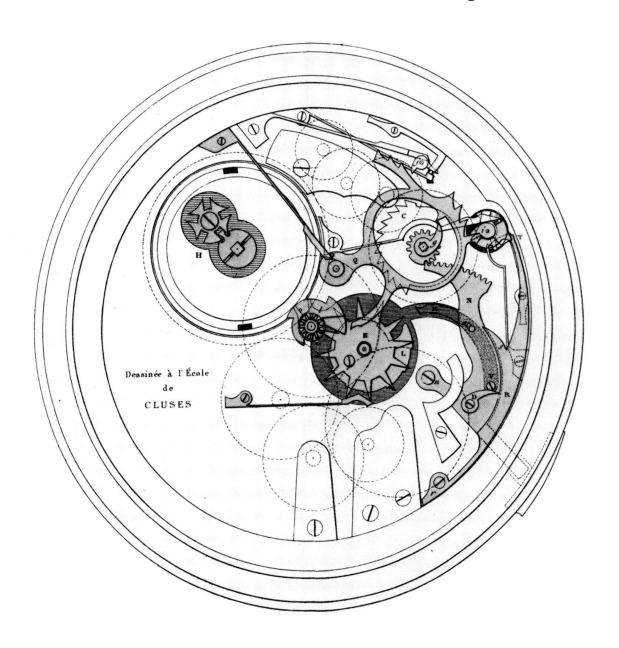

Dessinée à l'École
de
CLUSES

INTRODUCTION

The word 'clock' is used in a generic sense because its current meaning of 'timekeeping' is understood by anyone knowing the English language. It has been adopted by various of the sciences, such as biology, which has biological clocks, and astronomy with clock stars. Botanists call the seeding head of a dandelion a clock because its seeds are released at intervals by the wind.

The water clock was called a *horologium* by the Romans, and the word was adopted by the Latin-speaking scholars of Europe during the Roman conquests and after the fall of the Roman Empire. When the weight-driven mechanical clock was invented in the thirteenth century, the same word embraced both kinds of timekeeper. During that century and the fourteenth, it became corrupted in various languages to *horloge, orologio, relógio, reloj, orreloige, arloige, reloige, orloige, relorge, oreloige, aureloige, orloge, horreloige* and similar names until it emerged as *horloge* in the fifteenth century.

A distinguishing feature of the first mechanical clock was that it struck a bell at the hour or sounded an alarm bell at a predetermined time and probably did not show the time of day by a dial or hand, as did the water clock. A mechanical *horloge* that struck a bell gradually became known as a *clocca*, after a Latin word for bell, to distinguish it from an *horloge* without a bell. The spelling changed with usage, and as early as 1371 there are references in Latin to a *clok* and in English 'till itte be hegh none smytyn by ye clocke'.

When a dial and hand were added to the mechanical clock to show time of day, the mechanism had to be redesigned so that it

had two or three parts. The alarm and striking parts became subsidiary to the timekeeping. The timekeeping part was therefore given a new name, the 'watch'. The word came from the old meaning of a division of the night into periods of time – three by the Jews and four by the Romans and the Saxons. Samuel in the Bible contains a description of Saul and his people smiting the Ammonites 'in the morning watch'. A portable timepiece which could be carried on the person was thus a 'clock' if it struck the hours and a 'watch' if it did not. If it did both, it was a 'clock-watch'. In some other languages, German for example, there is no distinction between clock and watch.

Even students of horology, the science of timekeeping, now forget their plentiful jargon and call all except personal timekeepers

Above. The goddess Attemprance, appearing from a cloud, grasping a Gothic iron clock with going and striking trains, from a fourteenth-century French manuscript.

Left: An Austrian automata clock made at the beginning of the nineteenth century by Franz Vochenberger. On the central zone of the dial is a moving see-saw.

Right: One of George Cruickshank's prints of 1 May 1827.

Below: A wooden clock dated 1669, probably made in Davos, Switzerland. The great wheels have wooden pivots, but other wheels have steel pivots running in brass. The wooden escape wheel has steel teeth. The lower dial shows quarter hours.

clocks. All devices that have reasonably constant periods and are used as timekeepers have therefore been gathered here under the same name. They encompass such natural clocks as the sun and Earth (whose motions are indicated by the sundial), the moon, the human body's internal cycles, and also those of animals and plants. They include all artificial timekeepers such as water clocks, sandglasses, mechanical, electric and electronic clocks. The name 'clock' applies whatever the periodic signal it gives – by sounding a bell, firing a gun, flashing a light, silently turning a hand or displaying a number, giving pulsed signals like radio stars, charging and discharging a condenser, or making someone feel hungry every day at 12.30. The common factor is that the device functions as a clock.

The division between the natural and artificial clock is useful and of ancient origin. The best-known early book on practical clockmaking in English, published in 1696, was called *The Artificial Clockmaker*. It appeared at a time when there were especially strong links between astronomy and clockmaking, and a distinction between the art of dialling (designing sundials) and clockmaking was necessary.

The search for perfect timekeeping is one of the most inspiring stories of man's brilliance of mind and skill of hand. It takes us from the cave dweller in a primeval forest who relied upon the working of the biological clocks of the animals he hunted to trap them for food; to the ancient and skilled art of the shadow clock; to the priests who constructed the calendar and the astrologers who controlled monarchs by time predictions; to the long and close association of astronomers with clockmakers; to the intense effort to use clocks for navigation at sea; to the great cathedral clocks and models of the universe; to the domestic clock, the watch and the ultimate quartz crystal and atomic timekeepers that showed Earth itself to be a poor and erratic timekeeper.

One fact emerges with increasing clarity: the natural timekeepers that once dominated man's activities were kindly and forgiving masters. The artificial timekeeper which is now in almost complete control of industrialized man is a totalitarian taskmaster. In some situations today, the natural timekeeper has been almost entirely ignored – in factories without windows and a workforce on 24-hour shifts, for example. We are forcing plants to follow the artificial timekeeper so that they flower and fruit when our needs and our clocks say they should. The inevitable result is emotional stress and perhaps breakdown in man and loss of virility and death in plant. Fortunately, experience of travel in space has emphasized the importance of not subjugating the natural timekeeping of the human body to man's highly efficient artificial clocks. As a result, the astronaut is allowed to sleep when his body says so, even when there is no natural night and day and the clock-controlled computer provides his working schedule.

Man's natural biological clock rebels when he tries to impose on himself his slow, hard-won progress in timekeeping from the shadow pole stuck in the ground to the atomic clock that keeps time to one second in 3000 years.

Killing-Time —

Too much Time

Behind Time.

Trifling Time away —

Idling Time. away —

I
THE EARLIEST
CLOCKS

One of the earliest measurements of time was the calculation of 30 days between full moons. Ancient man depended on the moon as well as on the sun. It was his only general lighting at night. Some awakening intelligence must have noted that after he had made 12 notches on a stick marking full moons, the seasons were back to the point at which he had started to make his record. So he divided the year of seasons into 12 moons or months of 30 days each, which added up to 360 days in the year.

The Babylonians had a year of 360 days and there may be a connection between this fact, their division of a circle into 360 degrees, and their counting in 60s. Babylonian architects certainly knew how to divide a circle into approximately six sectors by setting a pair of dividers to the radius and stepping off this measurement around the circumference. But how and why did they divide each of these sectors in 60?

There is a big mental leap between dividing a year into 360 days and dividing a circle into 360 degrees or a day into hours and minutes. Days are discrete events; one is separated from the next by a period of darkness. The same is true of months and years. But to divide a day into hours or a circle into degrees involves the making of an artificial scale, because there is no natural division between hours or degrees.

There may be a connection between the hour and the division of a circle into six by dividers. It was easy enough to divide each division once more to obtain 12. To the Babylonian priests and architects, who were closely associated in the building of temples, it must have seemed a divine law

that the year was divided into 12 by the moon and that the circle was divided 'naturally' into six then 12 by the association of its radius with its circumference.

Perhaps that was how the day came to be divided into 12 hours. But, as well as the association with months in the year, there were undoubtedly mystical associations. All civilizations seem to have counted their

Far left : Seen from the centre, the sun rises over the heel stone at Stonehenge at midsummer. The stone ring was a calendar, indicating also the winter solstice and the equinoxes.

Left : A sun god tablet on which two lesser gods introduce King Nabu-apal-iddina (885–852 BC) of Babylon into the presence of the sun god Samos, sitting in his shrine. The Babylonian priests calculated a year of 360 days from the sun and seasons, and may have introduced the hour.

Left : The Aztecs made complex calendar calculations. This calendar stone, once vividly coloured, stood on a platform half way up the pyramid of the Aztec capital, Tenochtitlan. It is said to contain enough information to predict solar eclipses. At the centre is the sun god, Tonatiuh.

Right : King Akhnaton, Queen Nefertiti, and one of their daughters, adoring the sun in a Cairo Museum relief from Tell-el-Amarna (1552–1306 BC). The Sun was giver of life, provider of time.

shorter periods of time in units of 4 (the Saxon tides), 6, 12 or 24. The number 12 appears often in religion, myth and legend, as in the 12 labours of Hercules, the 12 great gods of Olympus and the 12 apostles.

When calculating the length of a year, it must have become evident many thousands of years ago that the figure of 360 days was not accurate and became out of step with the moons of 30 days. Some ancient astrologer, using a shadow stick, measured the point in the sky where the sun rose and noted that it moved northwards to a limit and then southwards to another limit and back again. The full cycle took 365 days. The priests thereupon set up permanent markers by which these important events could be noted and celebrated. One such marker is the large stone at Newgrange, 50 kilometres (30 miles) north of Dublin, built by the people of the late Stone Age in about 3100 BC. There is a tunnel 25 metres (80 feet) long, leading to a burial chamber. The tunnel was orientated very carefully when it was built so that a beam of light from the sun at sunrise on the shortest day shone along it. (The beam is a little off-centre today because of the gradual change in the Earth's inclination to the sun caused by the precession of the equinoxes.)

The great prehistoric monument at Stonehenge, probably built between 1900 and 1600 BC, is not just massive rings of arches. From the 'altar' stone in the centre of the ring, looking through the centre arch of those now standing, you can see the sun rising over the tip of the distant heel stone on the longest day of the year.

Left : Cleopatra's Needle, on the Thames Embankment in London, once stood at the temple of Heliopolos, a sun-worshipping centre in Egypt, where its shadow indicated the hour.

SHADOW RECKONING

The Greek historian Herodotus, writing in about 430 BC, ascribed the invention of the sundial to the Babylonians. Probably there was no single place where the sundial originated, but we know most about the sundials of Egypt, one of which is familiar even to Londoners. It is Cleopatra's Needle on the Victoria Embankment of the Thames, which was one of two calendrical monuments that, in about 1500 BC, stood outside a temple in Heliopolis, a centre of sun-worship and learning on the Nile delta. In the 1870s it was towed across the seas in a container to the Thames 'at the expense of a private individual . . . after having gone in its transit from the banks of the Nile at considerable risk of foundering in the Bay of Biscay', according to a contemporary report. The other obelisk from Heliopolis is now in Central Park, New York.

Another great obelisk, originally set up in the city of Thebes (now Luxor in Egypt), was also taken to the west by the Egyptologists, and it now stands in the Place Vendôme in Paris.

Long before modern Egyptologists explored the ancient cities and took home their prizes, the Roman emperors had helped themselves to about 20 Egyptian obelisk timekeepers to adorn Roman squares, where some still stand. Pliny tells us of an obelisk 30 metres (100 feet) high erected by a king of Rome on a hill outside the town to mark the place where magistrates were chosen and soldiers sworn in. When the Roman Republic was established in 510 BC, the obelisk was moved to the Field of Mars, where it still remains. According to Pliny,

Below : The persistence of the priests' connection with time and the calendar is illustrated by this Mexican manuscript of c. 1525–50, where they are shown making astronomical observations.

it once had a pavement around it with copper inserts on which there were marks to show the hours. An apple of gold was placed on top of the obelisk to indicate the tip of the shadow. The obelisk was also used for astronomical experiments by the Romans.

For an obelisk to show the hours, a semi-circle would have been marked around the base and divided into 12. The Egyptians started their period of a full day from midnight, so dawn occurred at different times. If they had started their hours at dawn, like the Moslems, their obelisk sundials would have needed movable scales. To determine dawn, the early Moslems used a bundle of mixed black and white threads. Daylight hours started when the threads could be distinguished from each other.

The Egyptians were ahead of their time in identifying night and day as part of the same phenomenon. Many countries measured them separately; Japan continued to do so into the nineteenth century. In the English language we still have no un-ambiguous word meaning a period of a day and a night. We have to say '24 hours', because a 'day' can mean either the full 24-hour period or just the time of daylight.

An observation made by the Egyptians was that the shadow of an obelisk when at its shortest always pointed in the same direction, no matter what the season of the year. This direction is what we now call a meridian, a line joining north and south. It was at the centre of the scale of hours around the obelisk, marked midday, or the hour of six on the Egyptian scale of hours. Architects used it for aligning buildings, particularly temples, north and south. Christian churches are also aligned in this way, from east to west.

The day when the shadow was longest at noon was the winter solstice: about 21 December. The day when the shadow was shortest was the summer solstice: about

21 June. In between these two days of the year it was found that the sun rises due east and due west near to 21 March and 23 September (using our present calendar). These days are called the equinoxes because then the lengths of day and night are equal. They had great importance in ancient ritual, especially the spring equinox, and the arrangement of many prehistoric stone monuments suggests that three monoliths were erected in the form of a V, one arm of the V pointing to the midsummer sun and the other to the midwinter sun. A line bisecting the V pointed to the equinoctial sun.

Excavations in Egypt have disclosed another unusual shadow clock, comprising a wall and a flight of steps. The shadow of the wall fell across the steps and the number of steps in shadow indicated the time. In the Second Book of Kings in the Old Testament there is a reference to such a dial:

And Hezekiah said unto Isaiah, What shall be the sign that the Lord will heal me, and that I shall go up unto the house of the Lord the third day?

And Isaiah said, This shall be the sign unto thee from the Lord, that the Lord will do the thing that he hath spoken: shall the shadow go forward ten steps, or go back ten steps?

And Hezekiah answered, It is a light thing for the shadow to decline ten steps: nay, but let the shadow return backward ten steps.

And Isaiah the prophet cried unto the Lord: and he brought the shadow ten steps backward, by which it had gone down on the dial of Ahaz.

There were also smaller sundials in the ancient world. The earliest one of which anything is known is Egyptian. A fragment of the original in the Neues Museum, Berlin, suggests that it was like a T with the top bent. In use, the sundial was placed with the cross-bar towards the east in the morning and towards the west during the afternoon. The shadow of the cross-bar falling across the stem shortened towards noon and lengthened afterwards, showing the hours on a scale.

From Egypt, sundials were introduced into Greece and then into Rome, becoming common in both communities. The earliest Greek sundial recorded is that of Anaximander of 546 BC, and the earliest brought to Rome seems to have been one placed near the Temple of Quirinus in about 290 BC.

As civilization developed, timekeeping on a vertical style or gnomon with a horizontal scale was found to be insufficiently accurate. If the length of shadow alone is taken into account, different scales have to be used at different times of the year, because the sun rises to different heights in the sky. If only the angle is measured, the lengths of hours change as the sun sweeps a bigger arc of the sky with the approach of summer. Various ingenious ways were devised to improve accuracy. Marcus Vitruvius Pollio, writing in about 25 BC, described 13 varieties of sundial in his *De Architectura*.

An accurate type of Roman sundial of which an example still survives is based on a hemispherical hollow in the top of a block of stone. From the centre of this hollow, a style rises like a miniature obelisk standing in a bowl, the top of the style being level with the edge of the bowl. Twelve vertical divisions marked the temporal hours of the Roman day, which were counted from dawn. As winter turned to summer, the sun rose higher in the sky and the tip of the shortening shadow moved down the bowl. A horizontal line inside the bowl marked the summer solstice and another nearer the top marked the shortest day or winter solstice. A central line indicated the equinoxes. The vertical lines were curved so that the tip of the shadow showed the temporal hour at that time of year. A temporal hour, being one-twelfth of a period of daylight, gradually becomes longer as the longest day approaches.

Berosus, the Chaldean astronomer, is said to have invented this kind of sundial, known as a hemicycle, in about 300 BC. Many versions were made with the part of the bowl representing the night hours cut away. A hemicycle made of limestone dis-

covered at Civita Lavinia, near Rome, was found to have lines engraved in it that are accurate for 42° north, the latitude of Rome. The style was missing, but it originally projected from the uppermost curved edge and was therefore horizontal and not vertical. Since the tip was in the same place, its shadow followed the same course. The Romans were quite accurate with their dialling, as the designing of sundials is called. Pliny complained of a sundial that was inaccurate for Rome because it had been cut at the latitude of Sicily.

The hemicycle, taking account of both the length and the angle of the shadow, was quite accurate. Any dials that measure the angle of a shadow – the position of the sun in the heavens – have to be sited so that the shadow line representing midday is on the

Below : The hemicycle was used for accurate timekeeping in ancient Rome. That shown would have had a horizontal style. It divided the period from sunrise to sunset into twelve equal temporal hours and indicated equinoxes and solstices.

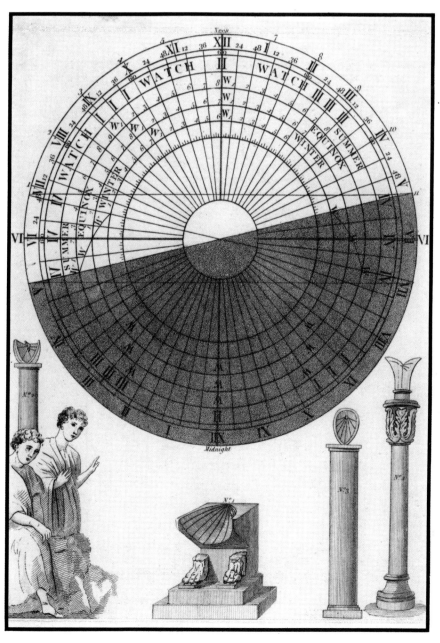

meridian, because this does not vary throughout the year. In other words, they have to be positioned accurately north to south, so most are fixed in position. After the compass was invented, many portable versions were made that incorporated a compass. They became known as compass dials and were made in large numbers in Europe during the Renaissance.

If only the length of the shadow is measured, the dial is known as an altitude dial because it measures the sun's height. All altitude dials have scales of dates and the date has to be known before the time can be estimated. Dials showing the angle and the length of shadow can act also as calendar dials and indicate the date.

The Arabs took over the art of dialling from the Greeks and Romans and they also

*Right: Dialling, the
art of designing
sundials, became a very
important church and
lay activity, and
eventually a hobby.
This design is from a
fourteenth-century
French Bible
commentary by
Nicholas de Cyra.*

*Left: A sundial is held
by a twelfth-century
stone angel on the south
porch of Chartres
Cathedral. The
original style was
horizontal and has
been replaced by an
angled one.*

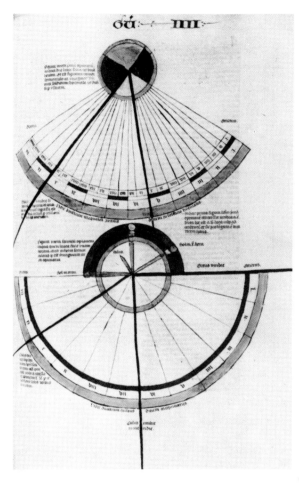

*Left: The earliest
known Islamic portable
altitude sundial, dated
AD 1159. It is
sideways here and is
held vertical by the
ring on the right to
face the sun. The
shadow of a peg in the
appropriate hole for the
time of year indicates
times of prayer and the
direction of Mecca.*

introduced another indication for religious
purposes. It was convenient to mark on a
compass sundial the direction of Mecca, so
that it would indicate to the faithful not only
the times of prayer, but also the direction
in which to pray. An Arab horizontal sundial of the fourteenth century AD, found at
Carthage, has marked on it the five prayer
times and the direction of Mecca.

Such indications are found only on horizontal dials. Vertical sundials were in the
form of columns. They were usually mounted
on the sides of buildings. A simple vertical
dial has a horizontal line with a rod projecting from the centre which is pointed towards
the midday sun. Underneath the rod is a
series of radiating lines to indicate the hours.
The Staatliche Museum in Berlin has an
Egyptian version with eight divisions marking the hours, which was made in the
thirteenth century BC.

Many examples of vertical dials are
scattered around the British countryside on
the south porches of ancient churches. They
can be recognized as several radiating lines
scratched or carved on a stone slab on the
south wall or on a pillar of the porch. Where
they meet there is a hole from which a
metal style once protruded. No such style
has survived, only traces of iron in the hole.
There is a fine vertical dial, which has been
dated between 1055 and 1064, just before
the invasion by the Normans, at Kirkdale
church near Helmsley in Yorkshire.

This Saxon dial is a semicircle with the
straight line at the top. Around the lower
part is an inscription meaning 'At every
tide'. There are four main divisions representing the four tides into which the
Saxons divided the daytime. The divisions
marking the tides have a short cross-line
across them. In between are shorter lines.
The makers recorded their names, 'Hawarth
made me and Brand priests', and added an
inscription that has been translated as 'Orm,
Gamal's son, bought St Gregory's minster
when it was all broken and fallen, and he had
it made new from the ground to Christ and

*Right: One of the
finest surviving dials in
Britain, dated c.1060,
is on Kirkdale Church,
Yorkshire. The four
divisions each side of
the vertical line
indicate the four
Saxon tides. The
horizontal style is
missing.*

St Gregory in Edward's days the king and in Tosti's days the earl'.

An earlier Saxon sundial, dated about AD 670, can be seen in the southern face of Bewcastle Cross in Cumbria. The divisions are similar to those in Kirkdale. The horizontal line at the top of the semicircle represents sunrise and sunset, and the central vertical line marks noon. Each quarter-circle is subdivided by lines at 45° to mark the two morning and two afternoon tides, and all the four shadow lines have the small cross-line at the ends. There are also divisions between the tides, but these may have been added later.

The Saxon's eight tides for day and night started at Morgen, 4.30 to 7.30 a.m., and went on to Daeg-mael, Mid-daeg, Ofanverth dagr, Mid-aften, Ondverth nott, Mid-niht

Right : A double sundial of 1836 on the parish church, Enschede, Holland. The figure of eight represents the equation of time. A spot of sunlight on the central line shows 'sundial' noon, and one on the curved line shows 'clock' noon.

and Ofanverth nott. All these were periods of three hours each. On the Kirkdale dial, the priests have their mark. The shadow line for Daeg-mael, which starts at about 7.30 am, is indicated by two short lines forming a cross and is thought to mark the start of a religious service. The Saxon three-hour divisions have survived in our church language of eventide and noontide.

In England, from the twelfth century to the fifteenth century, priests directed the villagers' attention to the times of religious services by means of vertical dials. Many such dials still exist on the south porches of churches of the time, in addition to the earlier Saxon dials. They do not mark the hours, but have only one or two shadow lines (and a hole for the style) to indicate the times of services. They are cruder than the Saxon dials, and are called scratch dials or Mass dials. As the local priest set the times of services, the dials vary considerably.

All the sundials of early periods described showed temporal hours, usually arrived at by dividing daylight into a number of hours, in most cases 12, and darkness into the same number of hours. In about the mid-fourteenth century, an Arab mathematician calculated that if the style of a sundial were parallel to the Earth's axis (that is, at the angle of latitude of the place where it was set up) it would show 'equal hours' – hours of equal length as if the whole period of night and day were divided into 24 at all seasons of the year.

This was of only academic interest at the time, since ordinary people, going about their tasks without benefit of artificial light, were only interested in the daylight hours

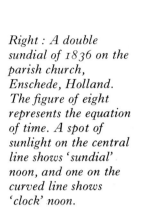

Left : A wooden cubical Florentine sundial for latitude 46° of c. 1560. Each side, when faced south, indicated hours in different systems including Italian, Babylonian, and Jewish.

Right : The oldest sundial with a gnomon (a cord) parallel to the Earth's axis. It is a compass dial of 1451 that belonged to Kaiser Friedrich III.

Far right : A shepherd's sundial, in use on the Pyrenees until recent times. It is an altitude dial, giving the hour by the length of shadow. The gnomon is rotated to the scale for the month.

for working, and temporal hours were adequate. Matters changed when the mechanical clock became more generally used, because clocks only registered equal hours unless they were modified in a rather cumbersome way to show temporal hours, as they were in Japan. The problem then arose of making sundials to show equal hours accurately, because the only means of setting a clock to time was by a sundial.

The earliest sundial known with a gnomon parallel to the Earth's axis is dated 1451 and is in the Landesmuseum Ferdinandeum in Innsbruck; it is a compass dial. There is a rare scratch dial showing equal hours at Litlington church in Sussex. All true compass sundials, including garden sundials, have gnomons angled parallel to the Earth's axis. It soon became evident to makers that if the gnomon were hinged, it could be angled against a scale marked in degrees and the sundial would become universal, suitable for any latitude.

However, it was discovered that, even with a sundial and a clock keeping equal hours, the sundial gradually drifted until it was faster and then slower than the clock in a cycle during the year. Much later, a few sundials were made that corrected the variation (the correction is known as the equation of time). One patented in 1892 by Major-General Oliver was followed by other versions. These are known as helio-chronometers.

The shadow clock has a history ten times longer than the mechanical clock and reached such a high degree of perfection that it took at least four centuries for the mechanical clock to displace it as a primary timekeeper.

THE WATER CLOCK

After defeating the Persians and the Egyptians in about 331 BC, Alexander the Great built a capital on the Nile delta that attracted all the learned men of the time. The town of Alexandria, with its libraries, reading-rooms and gardens, continued to attract men of genius after Alexander's death in 323 BC. Among these men were Euclid (*fl. c.* 300 BC) who laid down the theorems of geometry, and Archimedes (*c.* 287–212 BC), who explained the principles of the lever, the pulley wheel, the toothed wheel and the endless screw, as well as the fundamental laws of hydraulics.

One of those attracted to the engineering principles enunciated by the School of Alexandria was a barber called Ctesibius (*fl.* 2nd cent. BC). He applied hydraulics and mechanics to clocks and advanced the craft of the clockmaker by gearing the water clock to dials and hands and improving its accuracy.

One of his pupils, Hero of Alexandria (*fl.* 1st cent. AD), developed water clock-work to drive many kinds of toys. One of these was a model temple; the doors opened and the altar lit up, wine and milk were poured, drums and cymbals sounded, while Bacchanalian figures danced. The system devised by Hero is that employed today in automatic tea-makers.

The Roman historian Vitruvius later gave Ctesibius credit for inventing water clocks, but the Egyptians were using them in 1400 BC. One still exists and is in the Cairo Museum. It was found in fragments in 1904 in the Temple of Amon in Karnak. It is simply a bowl of alabaster with inscriptions on the outside and a hole near the bottom. It was probably filled with water at dusk and was used to show the night hours by the falling water-level inside. These were temporal hours, varying in length with the time of year. Around the inside of the bowl is a series of vertical dots, 12 rows in all, one for each month, and the names of the months are inscribed on the rim.

The clock is not as unsophisticated as it appears, however, because it has sides that slope outwards at an angle calculated to give the same rate of emptying whatever the level of the water. The angle of the sides is about 70°, which takes into account the effect of the viscosity of the water as well as the decreasing head. No doubt the angle was found by a process of trial and error. The outflow hole was made in precious metal or was a drilled gemstone cemented into the alabaster bowl to reduce wear. The function of the drilled gem was thus similar to that of the jewelled bearing in a wrist-watch today.

The Egyptians supplemented these out-flowing water clocks, which could be used in cascades, by inflowing water clocks. One was found at the Temple of Horus at Edfu and has been dated about 1300 to 1500 years later than the Karnak clock. The chamber is parallel-sided, presumably it held a float that rose as the water dripped in.

The Greeks used tapered bowls of the Karnak type for measuring the time of pleading in the courts, and one found in the Agora in Athens ran for six minutes.

The water clock is said to have been introduced into Greece from Egypt in about 400 BC by Plato, who is credited with the invention of a hydraulic organ clock that

Below left : Hero of Alexandria developed the hydraulic clockwork of Ctesibius to drive automata such as this 'miraculous altar', where lighting the fire caused the doors to open.

Below : Alabaster cast of an Egyptian water clock found at Karnak Temple and dated 1415–1308 BC. The level of the water, which leaks slowly from a hole at the bottom, indicates the hour.

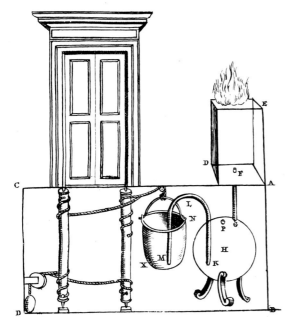

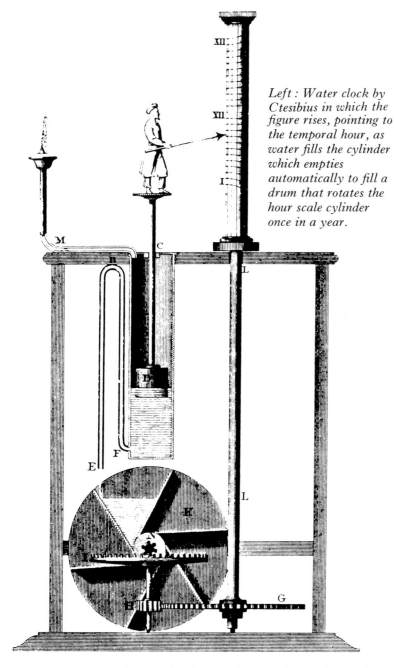

Left : Water clock by Ctesibius in which the figure rises, pointing to the temporal hour, as water fills the cylinder which empties automatically to fill a drum that rotates the hour scale cylinder once in a year.

The problem of showing temporal hours varying in length remained. Ctesibius overcame this by an ingenious automatic mechanism. A figure holding a pointer was raised by the float to indicate the hour on an adjoining column. The column was rotated once during a year by the water clockwork to offer an appropriate scale of hours to the pointer. A syphon on the side of the float chamber caused it to empty quickly when it was full. (The same system is used today in photographic print washers.) This water filled a chamber on a water-wheel which turned the cylinder by means of a series of gears.

Rome, the challenging power to Greece in the Mediterranean, undoubtedly had clepsydrae at an early time, long before the Roman architect Vitruvius was writing about them in the first century BC. According to Cicero, Pompey brought one back from the East after one of his campaigns. As it was an unusually accurate timekeeper, he used it to limit the speeches of the Roman orators. The Romans had copied the Greeks in timing lengths of pleas in court by using outflowing clepsydrae. Twelve clocks were normally allowed, but Pliny the Younger, when he was taking the case for the prosecution in a trial before the Senate in about AD 100, claimed four extra clocks because of the importance of the case. Since he recorded that he spoke for five hours, each clock must have lasted slightly over a quarter of an hour.

Julius Caesar is supposed to have come across a water clock in Britain after his invasion in 55 BC, because he made an observation with it that the summer nights were shorter in Britain than they were in Italy.

In ancient times in Ireland, a bowl with a hole in the bottom was used for time-telling, but in a different way from the Egyptian dribbling basin. It was floated on water until it sank. One found in a bog in County Antrim in Northern Ireland was made of bronze and took an hour to sink. It is thought that the sinking bowl was introduced by the Druids, who used it in Britain, Ireland and Gaul. The sinking bowl is not so much a clock as what is called today a timer. It measures an interval of time instead of indicating time of day, and was probably used for that purpose, like the sinking bowls of Algeria that measure the time a farmer is allowed to use water from an irrigation canal.

After the fall of Rome, Europe succumbed to the Dark Ages, when the light of learning

played the hours during the night and also showed them by a hand. A historical work, *Athenaeus*, described it as a round altar with the pipes pointing towards the water so that they produced a soft and pleasing sound.

One of the major problems with the water clock was maintaining accuracy. A clepsydra (water clock) with a hand and dial had a cylindrical float chamber that was slowly filled with water. As the float rose, it turned a hand by a rack and pinion or a weighted chain over a pulley. In order to ensure that the float chamber filled at a steady rate, it was fed through a funnel. At the edge of the funnel was an overflow, and water was fed into the funnel so that it overflowed continuously. The level therefore remained constant, so that the water ran into the clock at a constant rate.

Right : The Venerable Bede (c. 673–735), from a print dated about 1584. He wrote scientific studies of the calculation of time during the Dark Ages.

Far right : The earliest known illustration of a medieval horloge, in a manuscript in the Bodleian Library, Oxford, dating from about 1285. The hydraulic clock is depicted in the top circle and a Latin text beside it describes the miracle of King Hezekiah.

shorter'. The canonical hours indicated the times for chanting prayers. Originally there were only three, at approximately 9 a.m., noon and 3 p.m. Chanting itself was done to strict measure, and was a means of measuring the time of prayer.

The Benedictine Order, founded at Monte Cassino in Italy by St Benedict of Nursia in about 529, was especially renowned for learning and for the chanting of Gregorian plainsong. The Benedictines were, from the beginning, admirers of strict timekeeping. The monk Hildemar declared that 'no prayer is rational unless timed, by a clepsydra indicating the hours at night time or on a grey day', and the Roman Benedictine Cassiodorus commented that the *horologium* was invented for the utmost benefit of the human race.

There are many references in medieval writings in Latin to the *horologium*. In later English writings, the name became *horloge* or *horologe*. Similar derivations are found in other European languages. The Rule of the Cistercians, founded in 1098 to restore the Benedictine Order to its former austerity, directed the sacrist in these words: 'The sacrist shall set the *horologium* and cause it to sound in winter before lauds, on weekdays, unless it shall be daylight. And also to awaken himself before vigils every day. And after having arisen, he shall light up the dormitory and church. . . .'

Lauds and vespers were added to the canonical hours in the third century, making the total up to five. Another hour was added about 100 years later, and when St Benedict founded his order he tacked on one more hour, making the total seven. The canonical hours varied and differed in name in different orders at various times. A typical series runs: matins, lauds, prime, terce, sext, none, vespers and compline.

The sacrist or sacristan was responsible for the contents of the monastery, including robes and banners, gold and silver plate and holy vessels. The contents included the *horologium*. A sub-sacrist was usually given the duty of tolling the bell to call monks to service, for besides the *horologium* a monastery also had a bell that could be heard much farther afield than the clock. The Cistercian Rule of the early twelfth century refers to both. In the directions as to how the brothers should conduct themselves on weekdays in winter, it is stated: 'But after vigils, or after the office of the dead if this is due to be said, the brothers shall sit in the chapter house and those who wish

was almost extinguished. The spark that remained was blown into a flame in England by a Greek monk, Theodore (*c.* 602–690), who founded the School of Jarrow at the monastery there.

It was in this place that the Venerable Bede (*c.* 673–735) studied and wrote his historical books and scientific works which were mainly concerned with the measurement of time and the calculation of the calendar. Alcuin (*c.* 735–804), another monk, carried Bede's ideas to Europe, to the court of the Emperor Charlemagne. Charlemagne's palace became the centre of the intellectual revival that was spread abroad by schools he built on the Jarrow pattern in the closes of all great cathedrals and monasteries.

In medieval Europe almost all learning emanated from the religious institutions. Strict timekeeping was a vital element in most ancient religious orders and indeed remains so in those that survive. The abbeys had their own system of time, called canonical hours. The canonical hours were based on temporal hours, which, as St Augustine commented in his writings, had the disadvantage that, 'The hour in winter, compared with the hour in summer, is the may read . . . The *horologium* being heard, they shall attend to their necessities so that when the bell shall be rung, they shall be ready to go into the choir. . . .'

Elsewhere in the Rule the brothers are instructed that on a fast day in summer those going into the dormitory after sext (about midday) shall pause from their reading until the eighth hour (of daylight), 'And then the sacrist, aroused by the sound of the *horologium*, shall ring the bell as he is accustomed to do on other days'.

It seems that the *horologium* not only indicated the time but was an alarm as well. It would almost certainly have been a water clock, until about three-quarters of the way through the thirteenth century, when the mechanical clock was introduced. After that, whether the monastery clock was weight- or water-driven, it was still referred to as the *horologium* or *horloge*.

Recently an antiquarian, C. B. Drover, came across a remarkable picture in an illuminated bible dated about 1285. The manuscript is now in the Bodleian Library, Oxford, and the Latin text accompanying the illumination refers to the miracle of King Hezekiah and the dial of Ahaz. The picture shows a water clock. It is the only contemporary illustration known.

Although the working is not at all clear, there is no doubt about its being a water clock. Five bells are shown at the top and a sixth one in a lower corner. A circular dial with 15 divisions (it is thought that the artist made a mistake and there should have been 16) was probably rotated by the action of a float so that a pin placed in the dial at the appropriate place would trip an alarm. A bucket of water underneath may have been filled by the clock at the appropriate time to act as a weight and sound the alarm.

The manuscript is thought to have originated in northern France. Not far from this area, at Villers Abbey, near Brussels, some fragments of slate bearing instructions to the sacrist and his assistants were found at the end of the nineteenth century. They have been dated at 1267–8 and indicate how to set the *horologium* correctly for each day of the year, to allow for the varying periods of daylight and darkness as well as the differences in length of the church services because of church festivals. Directions for setting the clock according to the position of the sun on the church windows tell the clock-keeper, 'afterwards you shall pour water from the little pot that is there, into the reservoir until it reaches the prescribed

Above : A model of Su Sung's astronomical hydraulic clock tower of 1088 by Mr J. H. Combridge, in the Science Museum, London. The original tower was 11.3m (37 ft) high.

water, some to the well and others to the clock, while others with the utmost difficulty succeeded in extinguishing the fire with their cowls.

None of these fine old water clocks has survived in England, but some parts of one said to have been constructed as early as AD 125 still exist in the much less corrosive atmosphere of Greece. It is the monumental clock called the Horologion of Andronicus, which was in the Tower of the Winds, a picturesque tower standing at the northern foot of the Acropolis in Athens. Each of the eight sides of the marble tower represents one of the eight known winds. Once there was a pointer indicating the direction of the prevailing wind. On each side, under figures representing the winds, are engraved the lines of sundials so that the hours, the solstices and the equinoxes could be read. Of the elaborate clepsydra that was housed inside the tower, only various cavities, holes and pipes remain. The clock was fed with water from a reservoir in a round tower behind the south face of the Tower of the Winds. The clepsydra probably turned a hand to indicate the hours.

The Chinese seem to have developed the water clock to its most sophisticated point, probably between the seventh and the fourteenth centuries AD, and maybe from an even earlier century. The Chinese addiction to the keeping of accurate records has ensured the survival of a practical treatise entitled *New Design for a (Mechanized) Armillary (Sphere) and (Celestial) Globe*, written by a monk called Su Sung in 1090.

The descriptions and supporting illustrations are so comprehensible today that three models at least have been made of Su Sung's clock, which would be more accurately described as a great astronomical clock tower. It was over 9 metres (30 feet) high and topped by an enormous power-driven armillary sphere made of bronze for observation of the stars. An armillary sphere was an early astronomical instrument made up of bands representing the horizon, meridian, path of the sun and so on. Inside the tower a globe, also rotated automatically, showed the stars and where they might be seen from the observation platform above. The front of the tower comprised a five-storied pagoda, each storey having a door from which popped figures with bells to announce the hours and other special times.

The contraption was worked by a huge water-wheel which was restrained by an

level, and you must do the same when you set [the clock] after compline so that you may sleep soundly'.

According to the chronicle of Jocelin of Brakelond, in the year 1198 the water clock at Bury St Edmunds Abbey saved the relics of the saint. The wooden platform carrying the shrine and relics of St Edmund caught fire during the night. Fortunately, the clock sounded matins, waking the master of the vestry, who saw the fire and sounded the alarm. Jocelin recorded the excitement that followed:

And we all of us ran together and found the flames raging beyond belief and embracing the whole refectory and reaching up nearly to the beams of the church. So the young men among us ran to get ingenious device allowing the wheel to move in controlled jumps (a mechanism now known as an escapement). The water-wheel was released every quarter of an hour, which must have set the wooden tower and its mechanism shuddering and creaking.

Su Sung was staff supervisor of the Ministry of Finance in the Sung Empire, but he had studied the calendar in his youth. During his career he was sent on a difficult mission as ambassador to the barbarians in the north. It was at the time of the winter solstice, as predicted by the barbarians' calendar. The solstice was a day later by the Sung calendar he carried, which caused problems of protocol because the Emperor held the right to decide and promulgate the calendar. Astronomy was therefore 'nationalized' in medieval China. Indeed there were complaints at the time about astronomical instruments being too closely guarded by imperial officials. On his return, Su Sung reported privately to the Emperor that the barbarian astronomers' calendar was more correct than the Emperor's official one. The enlightened Emperor punished and fined all the officials of the Sung astronomical bureau. He ordered Su Sung to reconstruct the armillary clock

then in use. With a team of experts, Su Sung set up a working model in 1088 and completed the whole vast structure in 1094. Unfortunately, 20 years later a less learned group of barbarians from another area sacked the tower and took away the armillary sphere.

Much of the knowledge accumulated by the School of Alexandria, and subsequently developed by the Greeks and the Romans, was preserved by the wise men of Byzantium, which became the metropolis of the Eastern Roman Empire after the Western Empire was overrun by barbarians. Byzantium was renamed Constantinople in AD 330 and, in recent times, Istanbul. Here the Arabs and Persians learned the art of making timepieces and developed them in a new direction. The Arabs published several descriptive works on hydraulics, the earliest dated about AD 850.

One of the Arab water clocks was sent by the oriental potentate of Baghdad, Haroun al-Raschid, at the beginning of the ninth century to the Emperor Charlemagne. This was described much later, in the seventeenth century, as 'a machine driven by water showing the time, which is sounded by an appropriate number of small bronze

Below : General views of the inside and outside of Su Sung's clock tower, from woodcuts in an ancient text. The tower was built of heavy timber with two floors, one being the roof, where a clock-driven armillary sphere was sited under a pagoda. The first floor held the celestial sphere, rotated once a sidereal day, and the huge escapement was on the ground floor.

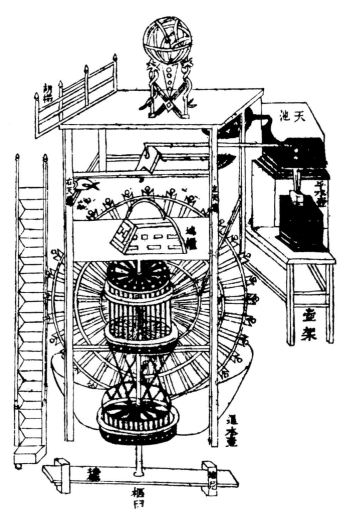

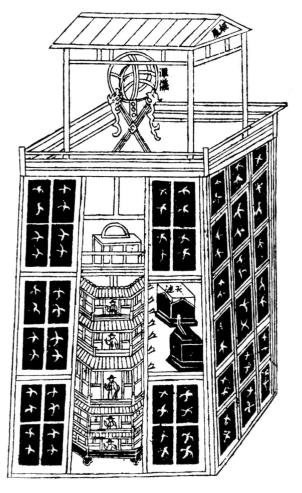

Left : A leaf from 'The Book of Knowledge of Ingenious Mechanical Devices' by Badi al Zaman ibn al Razzaz al-Jazari, showing a hydraulic automata which is an elephant clock. The manuscript, known as 'The Automata of al-Jazari', describes 50 devices under six headings, the first of which is clocks.

balls dropping into a brass basin, at noon, twelve horsemen come out of twelve windows, which close behind them. There are other wonderful things to admire about this clock, but it would take too long to describe them here'.

It is strange that although the Moslem religion forbade representation of the human or animal form, Moslems made and later imported many clocks with figures which were not only representative but animated. Water clocks that dropped balls into basins are known in a number of countries to which the Arabs penetrated. Part of one still exists at Fez in Morocco. It was built by the Emir Abou Inane in about 1357. Brackets of carved wood on an outside wall held the 12 bowls, and there is still a small part of the mechanism intact.

The man who built Haroun al-Raschid's clock, al-Jazari, also made a most remarkable clock in the form of an elephant bearing a mahout on its neck, a writer on its back, and a howdah with a third figure at a balcony in the front. The writer moved his pen to indicate the passing of time, while 15 little doors in the howdah opened one after the other. At the half-hour, a bird on top of the howdah turned and sang, and the man in the

Below : The first known illustration of a sandglass is in this Italian fresco of 1337–9 in the Palazzo Pubblico in Siena. To make it, two separate glass bulbs were bound together with a pierced diaphragm between them.

Above : Another illustration from 'The Automata', with some of the text. It shows the clock of doors, which has a candle at the centre and 14 doors around the candle holder. When an hour had passed from lighting the wick, a ball fell from a falcon's beak, the door opposite opened and a figure emerged. At each hour a different door opened.

howdah pointed to an eagle which dropped a ball into the mouth of a dragon. The dragon turned downwards to drop the ball into a vase. The mahout struck the elephant's head and the ball dropped on a cymbal to sound the half-hour. At the next half-hour, another eagle operated with another dragon.

All of this was worked by water, employing the principles set out in the first century AD by Hero of Alexandria in his book *Pneumatica*. There was a revival of interest in Hero's mechanisms in sixteenth- and seventeenth-century Renaissance Europe when hydraulic water gardens and grottoes

were built to astonish the visitor. Parts of some have survived, but the original mechanisms have disappeared. There were fountains, water gushing from animals, birds singing, organs and moving figures such as Neptune appearing out of the water, but no timekeeper.

The skill and the knowledge of the Arab clockmakers was admired in the West, but it could not be appreciated in detail until the last quarter of the thirteenth century when King Alfonso of Leon and Castile assembled a number of Jewish and Christian scholars and set them to translating into Castilian the learned works of the Arabic world. The result of their labours was a series of books of Moslem knowledge that included five on timekeepers: two books describe sundials, one a water clock, one a mercury clock and the fifth a candle clock.

THE CANDLE CLOCK

The candle clock was invented in England by King Alfred (849–99). According to his biographer, Asser, the King devoted eight hours of the day to public duties, eight hours to studying, sleeping and eating, and eight hours to worship. To apportion the time, he took 72 pennyweights of wax and

Right: A sandglass as used on sailing ships, hung from the cabin top by ropes. It marked the four-hour length of watches, about half the ship's crew being on duty during each watch. It was probably made in the eighteenth century.

Left: An eighteenth-century oil clock from north Germany with an egg-shaped glass reservoir mounted on a pewter stand. The reservoir is graduated from 8 p.m. to 7 a.m. to indicate the hours of night by the unburnt oil. It is inaccurate and was made partly as a novelty.

Right: The sandglass on the left of this illustration from a German treatise on the manufacture of fireworks, published c. 1450, was used as a timer for the stamping operation.

made 12 candles each a foot long and marked in inches. Each burned for four hours at the rate of 20 minutes an inch, the six candles lasting 24 hours. One of his chaplains, charged with looking after the candles, reported that they burned quicker in a draught, so the King devised a lantern or lanthorn with a frame of wood and sides of white horn scraped thin enough to be translucent, in which a candle could be mounted.

Some of the religious orders used candles in the same way to measure the canonical hours. Easter candles with bands marking the hours have persisted to present times. In Germany and France, oil-lamp clocks were in common use in the eighteenth century. A small glass reservoir for the oil was marked with divisions indicating the night hours as the oil was consumed.

THE SANDGLASS

The sandglass as a timekeeper is a relatively recent invention. The first known illustration is an Italian fresco of between 1337 and 1339, which is in the Palazzo Pubblico in Siena. It shows the early form of glass, with two separate bulbs bound together at the necks. A pierced diaphragm in the necks restricted the flow of sand.

A sandglass is fundamentally more accurate than a water clock, although they appear to be similar. If the conical shape of the bulbs is the same as the angle of repose of a pile of sand, the rate at which a sandglass runs out does not depend upon the height of the sand. Moreover, good-quality sandglasses, even centuries ago, were not filled with sand but with finely powdered eggshell, which ran more accurately.

Sandglasses were the industrial timers of the earliest factories. An illustration in *Das Feuerwerkpuch*, a German treatise on the manufacture of fireworks published in about 1450, shows an operation in a stamping mill being timed by a sandglass. They were also in common use at sea, because, if suspended, they operated well even in a moving and leaning sailing ship. And, of course, the clergy had good use for them. In sixteenth-century England, it was usual to stand a sandglass on the pulpit of a church to time the sermon. This was known as a sermon glass or a pulpit hourglass, and it was not uncommon for a loquacious cleric to announce 'Brethren, we will take another glass', and turn the sandglass upside down, to continue his sermon for another hour, as the sands ran down.

ᷓ̃nc ama
ui et equifi
ui a huue̅
tute mea
et queſini eam michi aſ
ſumere ſponſam. Ce
ſont les parolles que

ſalmon le ſage diſt en
ſon luure de ſapience
ou vin̅e chappitre. en
quoy il diſt. Iay ame
e ſapience et ſi lay que
ſe ſes en ma ieuneſſe
pur de elle faire mon

II
THE ADVENT OF CLOCKWORK

No one knows precisely when or by whom the mechanical clock was first invented. The little-known mercury clock, which was devised about the middle of the thirteenth century, had the essential elements of a mechanical clock. It was weight-driven: the rope turned a drum containing mercury. This drum turned at a uniform rate because it contained a number of vanes, each perforated. The mercury took time to run through the hole from one compartment to the next, controlling the rate of rotation of the drum. Thus the mechanism was an escapement. The drum drove an astrolabe dial, which is a movable star map rotated to show where the stars are at any time and day of the year.

This poses another question. Was the mechanical clock developed from the simple *horloge*, which sounded an alarm at a predetermined time or indicated the hour? Or was it developed by an astronomer as a means of indicating the passage of the stars? If we take the second view, the clock could have been 'naught but a fallen angel from the world of astronomy', as Dr D. J. Price, one of the translators of the Su Sung Chinese manuscript, put it.

CLOCKWORK

Gears or clockwork existed long before mechanical clocks. One of the earliest uses of gears was in the south-pointing chariot, in use in China perhaps as long ago as 2000 BC. A figure on the chariot was adjusted to point south. No matter in which direction the chariot was headed subsequently, the statue always turned to point to the south –

Opposite: In the fifteenth century, the clock became related to texts from 'The Wisdom of Solomon' and some were called 'Wisdom clocks'. The illustration is from 'Le Livre d'Horloge de Sapience' by Jehan de Souhande Velin, in the Bibliotheque Nationale, Paris.

Far left: The gear train of a calendar on an astrolabe made in Isfahan in 1221–2 by Muhammed Abi Bakr. It is the earliest geared device in a complete state extant, and is a forerunner of calendar work in mechanical clocks and watches. The triangular form of teeth had been in use since classical times.

Left: Back of the Persian astrolabe by Bakr, showing, at the top, the phase and age of the moon. Under it are two concentric rings each with a gold disc, showing the relative positions of the sun (the outer one) and the moon. The detachable back is unique.

obviously an extremely valuable aid when navigating overland. The mechanism was a form of differential gear connecting the figure with the axle of the chariot. It was a step towards the precision gearing necessary in clocks.

An earthenware mould has been discovered in China that had been used in the Han Dynasty (202 BC to AD 220) for casting ratchet wheels in bronze. It made wheels almost identical to those on spring-driven clocks of 1200 years later.

Both Vitruvius and Hero, the first about 100 BC and the other during the first century AD, described gearing. Almost all early forms were intended for the transmission of power, as from windmills and water-mills.

The Arab astronomers were possibly the first to make accurate trains of gears (with triangular teeth), for their astrolabes. One is described and illustrated in a text written by al-Battani in about AD 900, and a fine example that is very similar exists in the Museum of the History of Science at Oxford. It is Persian, is dated 1221–2, and was possibly a gift from Saladin to the Emperor Frederick II. 'The movable star map is on the front, and on the back is calendar information and a moon-phase indicator worked by the gearing. The moon is geared to pass through its phases in 29 and 30 days alternatively during the year.

THE MECHANICAL CLOCK

A treatise on elementary astronomy, written about 1271, gives a strong clue as to the time when the mechanical clock was invented. It was written by Robert the Englishman. Translated by Professor Lynn Thorndike, the relevant part of the text states:

Nor is it possible for any clock to follow the judgement of astronomy with complete accuracy. Yet clockmakers are trying to make a wheel which will make one complete revolution for every one of the equinoctial circle, but they cannot quite perfect their work. But if they could, it would be a really accurate *horologium* and worth more than an astrolabe or other astronomical instrument for reckoning the hours, if one knew how to do this according to the method aforesaid. The method of making such a clock would be this, that a man make a disc of uniform weight in every part so far as could possibly be done. Then a lead weight should be hung from the axis of that wheel and this weight would move that wheel so that it

would complete one revolution from sunrise to sunrise according to an approximately correct estimate. . . .

What held up the clockmakers of 1271 was the escapement. The 'disc of uniform weight in every part' was a flywheel, which is not as satisfactory as an oscillator for controlling the rate of a clock, although it has been used in a modified form. The problem had been solved earlier by Su Sung in his huge water-driven astronomical clock, but Western clockmakers at this time had no knowledge of Su Sung's escapement.

However, from the middle of the twelfth century, many Chinese mechanical ideas were transmitted to India and Islam and from there spread during the thirteenth and fourteenth centuries into Europe. One of these was the idea of using a lodestone to drive a model of the celestial sphere so that astrologers 'will be relieved of every kind of clock'.

But that was pie in the sky. The only method that proved feasible was to hold up the descending weight and allow it to drop a short distance at short intervals of time, in the same way as Su Sung checked his water-wheel.

Left : Front of the astrolabe shown on page 31. This is the rete side – a star map – in this case for 38 stars. Normally the rete is set to show star positions by an angle obtained by the alidade on the other side, but because there is a calendar on the back, two perforated projections are provided on this side for sighting.

A mechanical device which some antiquarians think was an escapement is illustrated in an album of architectural sketches prepared between 1240 and 1251 by Villard or Wilars of Honnecourt, near Cambrai. It was a weighted rope that checked a wheel and controlled an angel on the roof of a church. The angel pointed to the sun, which provided an indication of the time of day even when the sun could not be seen. Such an angel was once mounted on the roof of the cathedral at Chartres.

The clock of old St Paul's Cathedral also had a turning angel to indicate the time by pointing to the sun. A record for the year 1286 gives the allowances to Bartholomo the Orologius, the clock-keeper, as being one loaf daily. He was allowed 'bollae', too – presumably bottles of beer.

An indenture in Norman French between the Dean and Chapter and Walter the Orgoner of Southwark, dated 1344 and now in the British Museum, states:

> . . . The said Walter is to make a dial in the *horologe* with roofs and all kinds of housing appertaining to the said dial, and to the turning of the Angel on the top of the *horologe*, so that the said *horologe* shall be good, fitting and profitable, to show the hours of the day and night, to endure without failing . . . He is to receive six pounds sterling on completion of the work, is to find at his own cost the iron and the brass, etc., and is to have for himself the old apparatus which will no longer serve.

The mention of iron and brass indicates that the clock was mechanical. The reference to Walter's clock does not, however, mention a bell.

A translation of the *Chronicle* of Galvano de la Flamma refers to an iron clock being set up in a church in Milan in 1309. Dante mentions a clock in his *Paradiso*, written between 1317 and 1320, but it is not clear what kind; it could have been this striking clock or an alarm clock. He perhaps observed it in Milan where he was present at the coronation of Henry VII, the Holy Roman Emperor, in 1311.

Master Roger of Stoke was the chief clockmaker concerned with the building of an elaborate astronomical clock at Norwich Cathedral from 1321 to 1325. It had a great dial and 30 images, but only two jacks (moving figures) now remain.

While the first Norwich clock was under construction, Richard of Wallingford was at Merton College, Oxford, where there was a

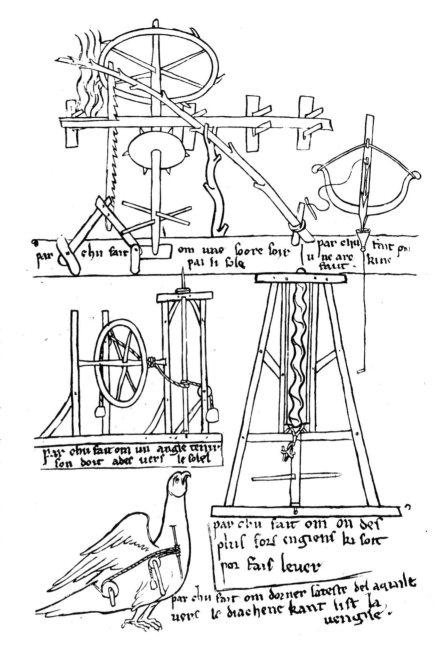

horologium. Richard left in 1327 to become Abbot of St Albans, and he subsequently produced many writings and instruments relating to astronomy, geometry and certain other sciences 'in which he exceeded all others at the time', according to the abbey's precentor about 50 years later. About the 1330s Richard of Wallingford designed an astronomical clock, which he had built and installed in St Albans Abbey. This clock is known to have incorporated oval gears and triangular teeth. Manuscripts at Gonville and Caius College, Cambridge, and a more detailed one in the Bodleian Library at Oxford, dated *circa* 1349–50, describe the division of wheels for an astronomical clock.

Some historians thought references to an 'albion' made before 1327 were to this

Above : An architectural drawing from an album made by Wilars de Honnecourt in 1240–51. The middle left drawing may have been a rope escapement to control the turning of a figure pointing at the sun.

astronomical clock, but the albion is now known to have been an equatorium, an instrument for calculating the solar and lunar longitudes and for predicting eclipses. It had sighting vanes and plumb bobs and incorporated two different types of astrolabe.

In 1364, Giovanni da Dondi (1318–89), who was professor of astronomy, logic and medicine at Padua University and professor of medicine also at Florence, wrote a detailed technical description of an astronomical clock that he had started work on in 1348 and finished in 1364. It showed the time in Italian hours, as well as giving many astronomical and calendrical indications. The instructions were so detailed that, in recent years, several reproductions of the clock have been made from them by Thwaites and Reed, English makers of turret clocks. The

Below : A fourteenth-century copy of 'Il Tractus Astarii', in which Giovanni da Dondi described in detail the construction of his planetary clock of 1348–64. The central drawing is by far the earliest of a clock escapement, showing the crown wheel and the balance, which is also like a crown.

original clock seems to have been destroyed during the Peninsular War, in a fire that burned down the Convent of San Yste in Spain, where it had been taken by the Emperor Charles V in the sixteenth century.

The most interesting aspect of Giovanni da Dondi's manuscript is that it contains the first drawing and details of a clock escapement. Infuriatingly, Dondi described the driving part of his astronomical machine as a 'common clock' with the 'usual' beat (rate of ticking). The drawing shows a horizontal balance wheel that looks like a crown of thorns. The teeth are triangular and the driving weights are shaped like blancmanges. Indeed, Peter Hayward, who made the first reproduction, used blancmange moulds to cast the weights.

The escapement is what is known as a crown wheel and verge. The crown wheel is rather like a band-saw. It has a flat rim with an odd number of teeth (pointed teeth like the points of a crown) at one side. The crown wheel is mounted vertically, and across the diameter of it is a vertical shaft or arbor on the top of which is the balance wheel. This arbor has two small flags or projections, called pallets, which intercept teeth on opposite sides of the crown wheel. The crown wheel is turned by the clock weight, but it cannot turn without its teeth rocking the pallets first one way and then the other from alternate sides of the wheel. The action of oscillating the balance wheel controls the rate of the clock – not very well, it is true, but much better than nothing and better than anything else that had so far been tried.

The vertical shaft with the two pallets and the balance wheel on top is called a verge, because it is like the staff with two flags that used to be carried by the verger in church ceremonies. Possibly the resemblance of the early balance wheel to a crown of thorns may have had a religious connotation as well. The priests still had a strong influence on time-keeping, because idle hands turned to mischief.

There is no clue as to who invented the verge escapement. The invention was so basic that the escapement lasted virtually without change for five centuries, from about 1275 to 1800. Although better escapements were devised after about 1670, only with the greatest reluctance were they eventually accepted by watchmakers.

An alternative to the plain balance wheel soon appeared. It was a bar with a weight at each end hung from one of a series of notches.

Above : The first reconstruction of Dondi's clock, made by Peter Hayward in 1961 for the Smithsonian Institution, Washington from the original instructions. It is seven-sided and stands about one metre high. The winged central dial is for mean time and the rising and setting of the sun. Above this is the Primum Mobile (sidereal time), with dials for Mars on the left and Venus on the right.

and is driven by a cord with a weight at one end and a smaller counterweight (to keep it in the groove of the driving wheel) at the other. A second wheel slows down the first and governs it by means of the foliot. The dial, he went on, is the daily wheel which makes a single turn in a natural day, just as the sun makes a single turn about the Earth in a natural day. The daily wheel carries 24 pins.

The striking part, said Froissart, is also complete with two wheels. The first wheel has a weight to drive it and is supported by a lever operated by the 24 pins. The second wheel is the striking wheel, which makes the little bells sound the hours by night and day, summer and winter. There must be a clock-keeper to attend diligently to the clock, to raise the weights, adjust the striking and set the foliot.

Early clocks, like Froissart's, sounded the hour by a single blow on a bell. What clockmakers call a locking plate or count wheel was invented at a very early date. This enabled a clock to strike the number of blows appropriate to the hour: a remarkable invention, because it gave the clock a memory.

De la Flamma's chronicles refer, under the date 1335, to a clock in the Church of the Beata Vergine in Milan as if it were unusual.

There is there a wonderful clock, because there is a very large clapper which strikes a bell twenty-four times according to the XXIV hours of the day and night, and thus at the first hour of the night gives one sound, at the second, two strokes, at the third, three . . . and so distinguishes one hour from another, which is of the greatest use to men of every degree.

The weights could be moved inwards to make the clock go faster and outwards to make it go slower. The fact that it was adjustable was a great advantage. Previously the only means of adjusting the timekeeping of clocks had been to increase or decrease the driving weight.

In 1369, Jean Froissart wrote a long poem, *Li Orloge Amoureus*, which compared the various parts of a clock with the attributes of love. It contains the first known description of this form of oscillator, which he called a foliot, the name still used. The French word *folier*, from which foliot is derived, means to dance about madly.

Froissart also described the other details of a contemporary clock in his poem. The first wheel, he said, is the mother and source of movement of the other parts of the clock,

The count wheel is a large wheel with a series of notches, at increasing distances apart, cut in its edge. When the striking train of gears is released at the hour (by a pin on a wheel of the going train lifting or dropping a lever), an L-shaped lever is lifted out of one of the notches of the count wheel. Each time this lever is allowed to drop on the 'land' between notches, a new blow is sounded, so the width of the land controls the number of blows (therefore, the widest land between notches represents twelve o'clock, and there is of course no land at one o'clock). The striking train also has a wheel with a series of pins in it to operate the bell hammer, and a rotating fan, known as a fly, to slow down the striking.

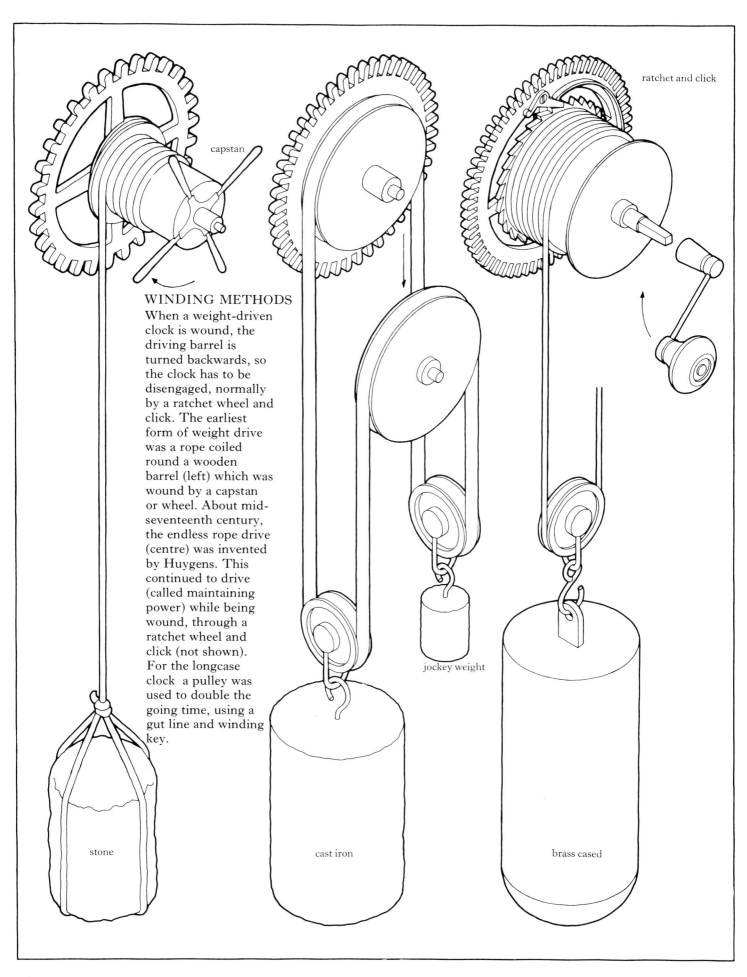

capstan

ratchet and click

WINDING METHODS

When a weight-driven clock is wound, the driving barrel is turned backwards, so the clock has to be disengaged, normally by a ratchet wheel and click. The earliest form of weight drive was a rope coiled round a wooden barrel (left) which was wound by a capstan or wheel. About mid-seventeenth century, the endless rope drive (centre) was invented by Huygens. This continued to drive (called maintaining power) while being wound, through a ratchet wheel and click (not shown). For the longcase clock a pulley was used to double the going time, using a gut line and winding key.

jockey weight

stone

cast iron

brass cased

These early clocks were quite large, from about 60 centimetres (2 feet) to 2.5 metres (8 feet) across. The frames were made of wrought iron with a post at each corner. Wrought iron was much used in the Gothic Middle Ages, and decorative grilles that have survived can still be seen in some churches and cathedrals. The clock frames were made by blacksmiths, who also made the great fire cradles for castles, from which the clock frame may have originated.

Many early clocks still exist, but few are as they were made. Most were converted to pendulum action after the pendulum was invented in 1657. That is understandable, but the wilful destruction in more recent years of parts of these old treasures is not. The oldest known clock striking quarter-hours is the city clock at Rouen in France,

Above : The clock in the church tower (1527–30) in the Piazza della Libertà, Udine, Italy, which has two jacks striking a bell alternately. The original wooden jacks were made about 1393, the present copper ones in 1850.

Left : One of the finest carved wooden jacks, a mid-sixteenth century ancient in armour at the Holy Trinity Church, Blythburgh, Suffolk. He stands 1.3m (4ft 4 ins) high.

built by Jehan de Félains in 1389. It was converted to pendulum in 1713, but kept most of its original wheels, the largest of which is about 1 metre (3¼ feet) in diameter. The frame is nearly 2 metres (6½ feet) wide and over 1.8 metres (6 feet) high.

There is another ancient clock at Dijon that may be as old as Rouen's. About the only original part remaining is the frame. It is magnificently made, but the rest has been altered and replaced. The clock came originally from Courtrai, but when Courtrai was pillaged after the battle of Roesbeke in 1382 the clock and hour bell were rescued by the Duke of Burgundy, who presented them to the town of Dijon, which had contributed 1000 men to the campaign. Some years later the clock was fitted with jacks to strike the bells.

Another French town, Montpellier, had an alarm clock to warn a watchman when to ring the big bell, but the watchman was so unreliable that they decided to sack him and order a new clock with jacks. This was in 1410, one of the earliest records of a man being displaced by automation.

A jack is an automaton, a mechanical doll that is animated by the clock to strike a bell, usually by wielding a hammer, but sometimes with his heels or elbows. It is still in use today for the roasting jack and for figures such as Jack-in-the-box and Jack-o'-lantern. Most early jacks were made of wood and represented men-at-arms. Either the whole figure pivoted to strike a bell with a hammer in the hand, or the arm lifted and was dropped. In the Wells Cathedral clock, Jack Blandifer's face and garments suggest

Below: A fifteenth-century Flemish illumination, showing a weight-driven tower clock through a window while King David is in bed and some attendants watch a battle raging outside.

a fourteenth-century origin, according to the late R. P. Howgrave-Graham, the expert on medieval clocks. The armoured jacks above the dial on the outside wall of the cathedral are definitely of the fifteenth century. At Southwold, in Suffolk, the jack is a colourful fifteenth-century warrior known as Jack the Smiter or Jack Southwold, whose goggling eyes, with red in the corners, and bluish chin are the original colours, the paint having been unintentionally preserved under a coating of bituminous paint.

MEDIEVAL TURRET CLOCKS

England is particularly rich in surviving medieval turret clocks. The oldest is that at Salisbury Cathedral. This was once converted to pendulum but has now been restored to its original foliot operation. It stands on the floor of the cathedral, and sounds the hours on a bell, which is the earliest known still in operation, separate from the clock.

Water clocks stood on the ground, and the tradition was continued with the first great mechanical clocks, which are illustrated in old documents as covered by heavily decorated wooden cases. The weights for the going and striking trains had to be suspended from pulleys higher than the clock, and, of course, tended to lift the clock off the ground. At a later date it became obvious that if the clock were in a tower, the weights could hang below it, keeping it steady; it could also be near the bell, and, if necessary, have an outside dial that could be seen from a long way off.

The Salisbury clock is wound by a three-spoked wheel with a rim of circular section, which is turned like the steering wheel of a truck. A ratchet and click on the barrel around which the rope is wound prevent the weight from falling. A similar wheel winds the striking weight. The original stones used as weights have long since gone and have been replaced by some of dressed stone. Stone weights were common, and there are still old church clocks which have their original weights: the one at Northend church, in Oxfordshire, for example. Winding wheels were often replaced later by a gear and cranked handle, particularly on the striking train, where the weight was somewhat heavier.

The Salisbury clock and its bell were originally set up in a bell tower that was detached from the cathedral and stood opposite the north porch. In 1790, James Wyatt, who 'restored' many church build-

Right : Part of the Salisbury Cathedral clock of 1386, working in the cathedral. The crown wheel with lantern pinion can be seen at about centre, with the verge and top pallet engaging a tooth. On the left is the striking train and a notch of the count wheel can just be seen. Bottom right is the winding wheel. The frame is held together by wedges and the trains are end-to-end.

Below : A continental turret clock (1737) by Martin Voght of Olomoue. Feet like this are not seen on English clocks. Although late, the frame is still wedged. The crown wheel is horizontal, at the top, with verge and pendulum. The trains are side by side, not end-to-end, and there are striking, going and quarter trains.

ings, pulled down two chapels, a porch and the bell tower at Salisbury. The old clock was moved to the cathedral's central tower. When a new movement was installed in 1884, the old one was left in the tower and attracted no particular interest until 1929, when T. R. Robinson drew attention to its great antiquity.

A search of the Salisbury Cathedral accounts showed the clock to have been made in or before 1386, because provision was made in that year for a house for the clock-keeper. Ralph Erghum was then Bishop of Salisbury. He was a regular visitor to the court of King Edward III, who had invited three craftsmen from the Low Countries to build clocks in England. They were Johannes and Williemus Vrieman and Johannes Lietuyt of Delft. Erghum engaged them to construct the Salisbury clock, and

39

Above: Dial of the clock at Rye in Sussex, with the 'quarter boys' above it. The movement dates from the fourteenth century and is much older than the dial. The pendulum is 5.5m (18 ft) long, beats 2¼ seconds, and hangs into the nave of the church to swing over the heads of the congregation.

when he was transferred to Wells, in 1388, they built another clock there. It was more up to date than the Salisbury clock because it struck the quarters as well as the hours. They probably also made the clock now in Rye.

In 1955 the Friends of Salisbury Cathedral asked R. P. Howgrave-Graham, who had authenticated Mr Robinson's recognition of the clock's antiquity, if it could be restored to its original working order. He and Mr Robinson agreed that it could be, and the firm of John Smith and Sons of Derby was entrusted with the task. After technical and antiquarian studies, Rolls-Royce X-rayed certain parts, to discover that some had been lengthened when the conversion to pendulum had been made. It was found that the clock had been converted twice to pendulum operation. A winding capstan had been fitted to the striking side to replace the winding wheel, and various other changes had been made. A new escape wheel and verge and foliot were made and fitted. These parts are painted a different colour from the original ones.

The earliest record of anything to do with the clock at Wells Cathedral in Somerset is in the accounts for 1392–3: 'In stipendim

custodientis la clokk 10s. per annum'. Later entries refer to some modifications and repairs. In 1480–1, the clock bell was changed and a 'wyndelas' was made for 'le Chyme'. Like the Salisbury clock, it has a massive wrought-iron frame with corner standards set at 45°, an angle similar to that of the buttresses of many early church towers, which the standards resemble also in their mouldings. Spokes of the toothed wheels are lapped over both sides of the rim and forged to it. These engage, not with other toothed wheels, but with lantern pinions, which are a series of bars between end-plates, like squirrel cages. This grand type of tower or turret clock was made until the early sixteenth century.

The Wells clock has an astronomical dial inside the cathedral and a simpler time of day dial outside, and it also operates jacks. The original movement was removed in 1835 and is now in the Science Museum in London. The second movement ran for 45 years and was then removed to the parish church at Burnham-on-Sea, where it still goes, and the third is of modern construction and is still at Wells. Unfortunately, all the connecting mechanisms to the dials were removed and thrown away during the changes. As may be seen, clocks last a very long time with little wear, even though they work continuously, day and night.

The large astronomical dial at Wells is one of the finest in the world, and illustrates the medieval concept of the universe. Around the outside are 24 markers, 1 to 12 repeated, against which a moving sun shows the hour. Inside is a minute ring with a star as the pointer. Inside this, the moon turns to show its age and phase, and in the centre is the fixed Earth. The movements of the sun and the moon correspond with those of the real sun and moon outside.

The hours are struck by the large and colourful Jack Blandifer, made of oak, who sits in a sentry box a short distance from the clock. He strikes the hours on a bell in front of him, using a hammer held in one hand, while he turns his head to listen. At the quarter-hours, he kicks bells with his heels.

There is another performance at the hours. Four knights on horseback fight a tournament in a Gothic turret above the astronomical dial. Two charge around in one direction and two in the other. One unlucky knight is felled every time by his opponent's lance and crashes back on his horse; he is uprighted behind the scenes, only to be knocked flat the next time round. When the clock was made, knights were still jousting in tournaments. The display is so popular with visitors that it is turned on at the quarters as well during the summer, but the long-suffering clock is then helped by an electric motor.

The outside dial is on the north wall and is marked from 1 to 12. Originally it had only an hour hand, but a minute hand was added at a later date. Hours are struck on a large bell on the cathedral's central tower, but cannot be heard from here. There are, however, above the dial two jacks in full armour who swing around to sound the quarter-hours on two small bells. There is probably no other clock in the world that strikes and chimes in two different places at the same time.

The going and striking trains were originally separate, and are also separate in the present clock. The hour bell is in the north tower, the home also of the striking train, which is released by the going train behind the astronomical dial in the cathedral. The quarters are unusual because they are struck on one bell, one blow for the first quarter, two for the half-hour, and so on. Most early clocks have two bells for the quarters,

Left : The old Dover Castle clock, one of the few remaining that retain their original foliots, seen at the top. Although usually dated later, it could have been made in the sixteenth century. The count wheel is in the centre, a design that originated about 1390. A number of similar clocks are known in the south of England. The wheel on the left is for winding.

Right : The wooden-framed turret clock of c. 1485 at Cothele House, Cornwall. Several of the type are known, the more northern ones with iron frames. The underslung foliot can be seen at the bottom.

Far right : Dial of the seventeenth-century astronomical clock at Wimborne Minster, Dorset, with Earth at the centre of the universe and the sun showing the time.

Below : The late fifteenth-century astronomical dial in the nave of Exeter Cathedral, Devon, with Earth at the centre. It is operated by a modern movement, but the original one is preserved in the cathedral. It comprises three parts of different dates, a four-poster of mid-seventeenth century, a vertical frame striking train probably of the sixteenth century, and an old quarter striking train on top.

known as ting-tang chimes because one bell has a higher note than the other. The first quarter is sounded as ting-tang, the second as two ting-tangs, and so on until there are four before the hour is struck.

Another clock in this group is that in St Mary's Church, in the historic and picturesque town of Rye in Sussex. It was probably made by one or more of the three Dutch 'orologiers' who constructed the Salisbury and Wells clocks, as the frame is very similar. Those wheels and pinions that are original are also almost identical. An examination of the punch marks used to mark out the wheel teeth shows that the same punch was used on the three clocks.

The Rye church authorities bought the clock in 1560, and it has been assumed to date from then, but T. R. Robinson made

a study of it in 1974 and came to the conclusion that it was of fourteenth-century origin. Lewys Billiard of Winchelsea, who supplied the clock, probably got it from a monastery, the Dissolution of the Monasteries being then a recent event, and restored it, perhaps to sell it to Rye as a new one. As Mr Robinson remarked, 'Even if they had known that their clock was from a monastery, the good church folk of Rye might have considered it inadvisable to admit that they had acquired seized church property!'

The size of the clock suggests that it came from an important place – Battle Abbey, perhaps. There have been substantial alterations to the clock during its life, including conversion to a long pendulum which hangs down into the nave of the church and swings over the heads of the congregation. The pendulum rod is stayed like a ship's mast. Two jacks sound the quarters, but they are made of fibre-glass because the old wooden ones rotted away.

The wrought-iron frames of the earliest clocks were held together by iron wedges driven into rectangular slots, like early wooden furniture. The screw and bolt did not appear until about 1500. The going train was mounted in one side of the frame and the striking train in the other, with the count wheel outside the frame at this end.

THE DOVER CASTLE GROUP

The next group of clocks is known as the Dover Castle group, and has the count wheel in the middle of the frame between the two trains placed end to end.

A manuscript of about 1390, entitled 'Une Petite Traité pour Faire Horloges', now in the Vatican Library (part of Codex Vaticanus Latinus 3127), describes the making of these clocks with central count wheels. Several are known, and there are probably others that have not yet been recognized for what they are. They are named after an example from Dover, now in the Science Museum in London, that has its original escapement. The British Museum also has one with original foliot that came from Cassiobury Park.

These clocks are smaller than the Salisbury group, being less than 60 centimetres (2 feet) wide against over 1.5 metres (5 feet). There are others at Clandon Park, near Guildford, in Surrey; at St Peter's Church, Buntingford, Hertfordshire; and in a stable block at Quickswood Farm, Hertfordshire. Similar clocks are also to be found in Kent, in

Chilham, Charing, Eynsham, Wingham and Wrotham, and there are others in Surrey, Bedfordshire, Sussex and Cambridgeshire. Some have a maker's mark like a cockleshell.

It is probable that the designs of the Dover Castle-style clocks – possibly some actual clocks – were imported from Holland or Belgium between 1490 and 1500. The distribution suggests that the designs and possibly clocks were imported into the south of England.

When the English started to design and make clocks they did not slavishly copy those made by workmen from Europe. They used vertical frames of iron or wood. The first English-made clocks were large, standing as high as or higher than a man. The wheels were mounted in a line and the going train was 'upside down', with the foliot at the bottom instead of the top.

Most English vertical frame clocks of this period appear to have been made either in the West Country, where the frames were of iron, or in the Midlands, where wooden frames were preferred. One with an iron frame, and its verge and foliot mechanism intact, can be seen in the National Trust property Cothele House, in Calstock, Cornwall.

Perhaps the most famous of this group of fifteenth- to sixteenth-century turret clocks is that at Exeter Cathedral, which replaced an earlier one of about 1384. The going side of the clock is of vertical iron-frame construction, but not the striking side which is of a birdcage design of the seventeenth century.

The Exeter astronomical dial is operated by a movement of modern construction (a phrase that, in clock terms, means that it was made in the last hundred years). The frame of the clock is of flat-bed design, being two girders with the wheel arbors (axles) in a row.

The Exeter Cathedral astronomical dial is less elaborate than that at Wells. However, as at Wells, the Earth lies at the centre of the universe. Around the Earth is a ring of numbers showing the age of the moon, which is itself represented by a globe, half white and half black, rotated to show the phase. The outer ring of numbers is a double-XII, against which the hours are indicated by a red disc representing the sun. The Exeter clock originally had no minutes, but a separate large minute dial with a hand was mounted above it at a later date.

There is another clock with an astronomical dial in the church at Ottery St Mary,

Above : The astronomical clock and automata at the Hotel de Ville, Prague. The original clock was built about 1486. Below is a calendar dial, and above, at the top of the tower, more normal dials which are driven by a separate movement.

Left : A small tower by Chartres Cathedral houses this 24-hour dial. The many rayed sun is carved from one piece of stone 2.5m (8 ft 6 ins) in diameter.

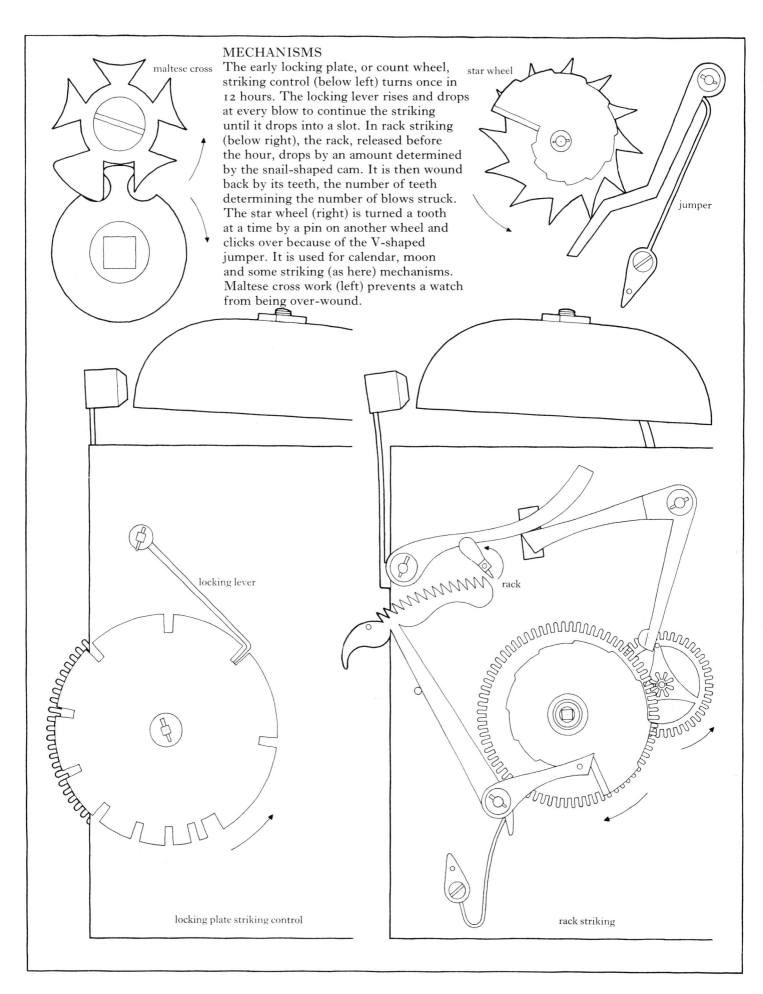

maltese cross

star wheel

MECHANISMS

The early locking plate, or count wheel, striking control (below left) turns once in 12 hours. The locking lever rises and drops at every blow to continue the striking until it drops into a slot. In rack striking (below right), the rack, released before the hour, drops by an amount determined by the snail-shaped cam. It is then wound back by its teeth, the number of teeth determining the number of blows struck. The star wheel (right) is turned a tooth at a time by a pin on another wheel and clicks over because of the V-shaped jumper. It is used for calendar, moon and some striking (as here) mechanisms. Maltese cross work (left) prevents a watch from being over-wound.

jumper

locking lever

rack

locking plate striking control

rack striking

Devon. The early seventeenth-century Ottery St Mary clock has the same general layout and globular moon as the Exeter clock, but the hours are indicated by a star-shaped pointer opposite a sun that also goes around the dial. It is probably the only clock in this group that has its original movement. This is exposed and can be seen in the fine clock gallery in the south transept, working its original astronomical dial, although the stars with which it was originally decorated have been painted out.

Astronomical dials appear to have been simplified, probably because of cost, over the years. There is a later, and again simpler, early seventeenth-century astronomical dial at Wimborne Minster, in Dorset. Again there is a fixed Earth in the middle, but there is no indication of the age of the moon, although the moon is shown as a rotating globe in a fixed position. On the outside, a sun turns to indicate the hour on a double-XII dial. The numerals are on little shields mounted at an angle. Unfortunately, someone has stuck two Renaissance angels and a quite incongruous urn on the top.

Clocks with geocentric astronomical dials and double-XII numbering similar to those in England are found in a number of places on the Continent, including Chartres, in France; Sion and Soleure in Switzerland; Lund in Sweden; and in Germany.

THE STRASBOURG CLOCKS

Europe has a wealth of astronomical clocks. The most famous and elaborate of them all is in Strasbourg Cathedral. The first Strasbourg clock was built in 1352–4, and the second was built in 1574 by the famous clockmaking family of Habrecht. The work was supervised by Conrad Hasenfratz (known as Dasipodius, the Latin translation of his name, meaning 'hare's foot'), who was professor of mathematics at Strasbourg University; he also calculated the astronomical trains. This monumental clock, with numerous automata as well as multiple dials, was 10.4 metres (34 feet) high.

At the end of the eighteenth century, the Strasbourg clock had stopped working and no one could be found who was capable of repairing it. There is a story that a boy of ten was listening to the tale of how the great clock used to work and resolved there and then that he would be the one to restore it. His name was Jean Baptiste Schwilgué, and he later taught himself mechanics and horology so successfully that he became professor of mathematics at the College of

Silestat. He set up a business with a partner to make turret clocks and weighing scales, but never forgot his childhood resolution. As his fame grew, the Mayor of Strasbourg invited him to give an estimate for the repair of the great clock. Schwilgué made three proposals, one of which was that the old clock should be replaced by a new one. This suggestion was accepted, but the new clock was not completed until 1842.

In front of the clock is a celestial sphere that turns to indicate sidereal time and also shows the precession of the equinoxes. At the lower level the clock has a complicated calendar which automatically shows all the moveable feasts, such as Easter. At this level there are also some complex astronomical indications.

Above the annual calendar in the centre is an ordinary clock dial with on each side a model of a small boy, one striking the hours and the other turning a sandglass every quarter of an hour. Higher again is a planetary dial with signs of the zodiac showing movements of the planets accurately; and above that a globe turning to give the moon's phases.

Near the top is a series of automata or jacks showing processions of the ages of man and the apostles passing before Christ. On top is a metal cockerel that moves and stretches its wings as it crows at 12 o'clock.

The original clock mechanism and some relics from the first clock of 1354 and much of the dial and most of the movement of the second clock of 1574 are exhibited in the Horological Museum at Strasbourg.

LATER TURRET CLOCKS

Early turret clocks were controlled by foliots with verge escapements. After the pendulum was invented in 1657, many were converted to pendulum operation, with a new, differently placed, verge. About 1670 a form of escapement called the anchor was devised that allowed the use of a longer pendulum and gave it more dominance over the clock. Most earlier clocks were altered to this system, and nearly all new ones were based on it.

In France, a much-simplified frame had been pioneered, the flat-bed. It was strong and simple, and had the great advantage that wheels and their arbors could be removed and replaced without completely dismantling the frame. This pattern was adopted by Edmund Beckett Denison when he designed Big Ben. After 1858 this set the model for other turret clocks.

46

III
DOMESTIC CLOCKS

The domestic clock was not exactly invented. It was probably a spin-off, to use a modern term, from the scientific activities of ecclesiastics, astrologers (astronomers) and mechanicians of the Middle Ages interested in increasing their knowledge of the stars or improving discipline in religious communities. History was repeated in the twentieth century. The electronic watch was a spin-off from space and computer technology, because the money that financed the first American moon probes paid for the development of a time switch that became the module of the first electronic watch in production.

Maybe some thirteenth-century king or prelate first had a clock in his house as a symbol of prestige or wealth, or perhaps from interest, or to call him to prayer. Certainly the sexton's alarm to warn him when to ring the monastery bell became the watchman's clock of the lay community, for the watchman's duties were similar to those of the sexton. He had to ring a bell in the watch tower to warn the villagers of curfew, when fires had to be put out, and when they were needed for some communal activity such as ditch-digging or repelling invaders.

So perhaps it was the watchman's clock, hanging on the wall, that became the domestic iron clock of the medieval household. It was a valuable possession and when the family travelled it went with them (as did the glass windows). A later domestic or monastery clock, a seventeenth-century lantern clock, which was probably being carried around like this, or may have been stolen, was found by skin' divers of Essex Underwater Research in 1973.

Iron clocks and lantern clocks, hanging on the wall from a hook and prevented from moving by two sharp spurs at the back of the clock, were the first general domestic clocks. The weights hung below them and generally had to be pulled up twice a day. In some countries, it became fashionable to fit ornate wooden cases around them and mount them on wooden brackets.

Far left : A tabernacle clock 52cm (20½ ins) high of the first half of the fifteenth century, made in Augsburg. The case is gilded base metal and the pendulum, which hangs in front of the dial, was added over a century later.

Left : An iron chamber clock is shown in some detail in a French manuscript of 1454 in the Bodleian Library, Oxford. It stands on Earth with its movement and bell in Heaven. The goddess Attemprance has one hand, incongruously, in the wheelwork.

was most useful for driving the clock, but the reducing force was a problem.

Early coiled springs also suffered from clustering, that is, becoming partly coil-bound from time to time, because they could not be made very evenly or smoothly and did not coil accurately. When this happened the power was released in uneven bursts. The means adopted to overcome these disadvantages, which directly affected time-keeping, were twofold.

The first step was to limit the use of the spring to the middle of its action, by using stopwork to prevent it from driving the clock when it was over- or under-wound.

The next step was to provide a form of gearing between the spring and the clock to make the power output even. One method used for some of the earliest watches was to dissipate some of the mainspring's power by a form of brake known as a stackfreed, but by far the best method was an early solution. It was so simple, so ingenious, so elegant, that it has remained in use, at least in certain special clocks and chronometers, from the time it was invented, in the early fifteenth century, until today.

It is called the fusee, probably from the Latin word *fusata*, meaning a spindle wound

Left : The earliest known spring-driven clock belongs to the Victoria and Albert Museum and is on loan to the Science Museum, London. It is similar to weight-driven clocks of the time, but has spring barrels in the base instead of weight barrels. The dial is missing.

THE SPRING-DRIVEN CLOCK

Although the weight-driven clock was not invented for domestic use, the spring-driven one undoubtedly was. The use of a coiled spring instead of a weight to provide power made possible first the portable clock and subsequently the personal clock, now called the watch. Spring clocks were being made in Burgundy about 1430, it seems, but little is known of their origin.

The earliest spring-driven clock known is dated about 1450 and belongs to the Victoria and Albert Museum in London, and is on loan to the British Museum. It is like the weight-driven chamber clock of the time, which was fastened to a wall and had the weights hanging from winding barrels at the bottom of the clock. Instead of the weight barrels there are spring barrels — drums with coiled springs inside them.

The change-over did not prove as simple as that, however, because the coiled spring, made of hammered brass, did not provide a constant source of power, as the falling weight did. When wound up it gave a force that was very strong, but only for a short time. The force was then medium and decreased unevenly for some hours before falling off rapidly. The middle of the range

Left : The earliest domestic clocks were made of wrought iron and were without cases. This example was made in the fifteenth century in south Germany and has an oscillating foliot at the top. It is weight-driven and the hour was struck on a bell on top as well as being shown by the single hand.

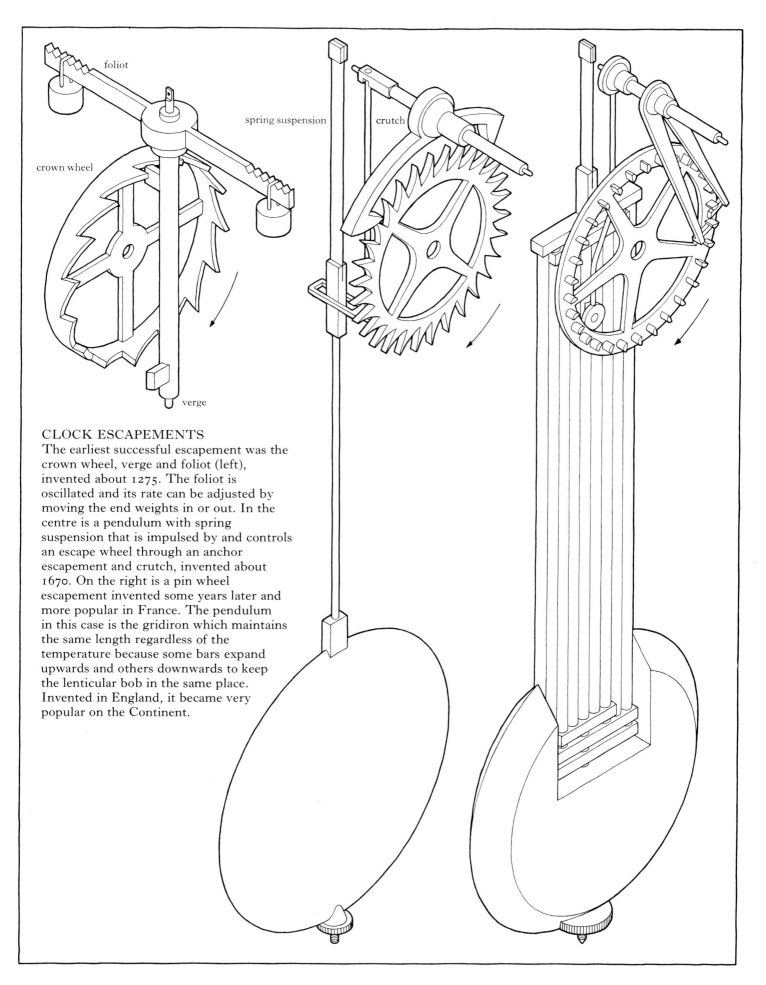

foliot

spring suspension

crutch

crown wheel

verge

CLOCK ESCAPEMENTS

The earliest successful escapement was the
crown wheel, verge and foliot (left),
invented about 1275. The foliot is
oscillated and its rate can be adjusted by
moving the end weights in or out. In the
centre is a pendulum with spring
suspension that is impulsed by and controls
an escape wheel through an anchor
escapement and crutch, invented about
1670. On the right is a pin wheel
escapement invented some years later and
more popular in France. The pendulum
in this case is the gridiron which maintains
the same length regardless of the
temperature because some bars expand
upwards and others downwards to keep
the lenticular bob in the same place.
Invented in England, it became very
popular on the Continent.

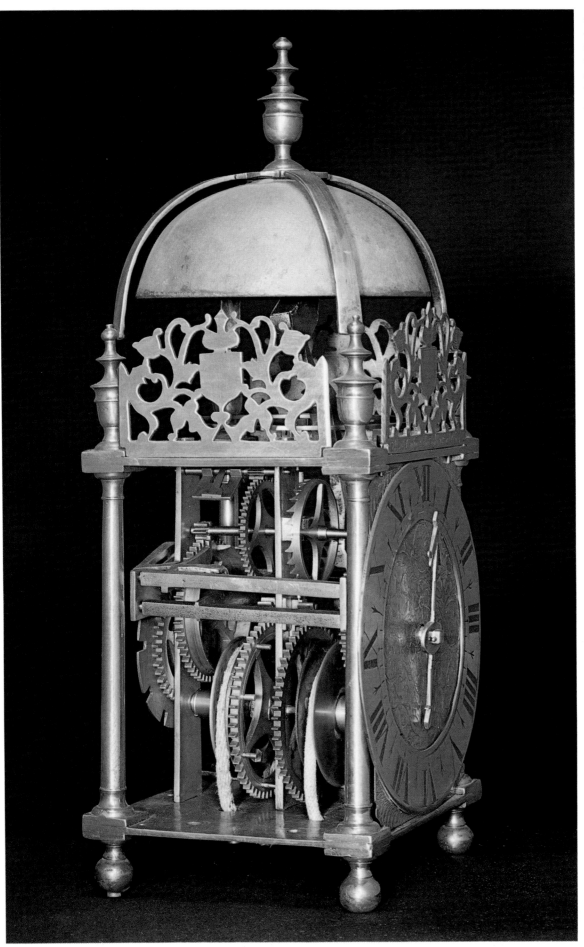

Left : The earliest English-made and designed domestic clocks were brass lantern clocks. This one was made by Thomas Knifton, Cross Keys, Lothbury in about 1650. It is weight-driven, strikes the number of hours, and indicates them by one hand. The verge escapement has a circular balance without a balance spring under the bell.

Right : A spring-driven table clock of the belfry style preferred by Flemish and Burgundian makers. The Frenchman Nicolas Lemaindre made this one about 1619 for Marie de Medici, second wife of Henry IV. The fusee can be seen clearly in the going train at the bottom ; the striking train is above it. The bottom plate on the left has elaborately pierced and engraved bridges and cocks and a circular balance.

with thread. A fusee is a trumpet-shaped device with a toothed wheel at the larger end which engages with the first driving wheel of the clock. The trumpet-shaped part has a spiral groove cut in it, and a length of catgut is attached to the groove at the larger end. The rest of the gut line is wound around the barrel containing the spring.

When the fusee is turned with a key, the gut is pulled off the barrel, which winds up the spring inside it. The gut is wound on the fusee groove, which becomes smaller and smaller in diameter. There is a ratchet arrangement to prevent the clock wheels from being turned backwards while the fusee is being wound.

When winding is complete, the spring turns the barrel in the opposite direction, which pulls the gut line and turns the fusee to drive the clock. As the spring runs down, however, the gut is pulled off an ever-increasing diameter of the fusee groove, so increasing the leverage or mechanical advantage and keeping the torque applied to the clock more or less constant as that of the barrel lessens. To put it in another way, the spring barrel is allowed to turn faster and faster to provide the same power output.

The taper of fusees changed over the years as springs were made stronger. The earliest known illustration of a fusee in a clock is in an illustrated manuscript of 1450–60 in the Royal Library in Brussels, which shows a seated cleric and a standing woman, both holding books, with two very large standing clocks, several sundials, and the movement of a table clock in which the fusee is clearly depicted. Leonardo da Vinci also sketched a fusee in his manuscripts of 1485–90.

The first printed illustration of a clock movement, which included a fusee, appeared in one of a series of 17 books 'On Divers Things', first printed in Basle in 1557. The author was Jerome Cardan, an Italian physician and mathematician, whose name has been perpetuated in the cardan shaft and joint used in the transmission of a motor vehicle.

This illustration omits the vital element of the escapement, and Cardan's description shows that he had no idea of what the fusee did. He added a note about the spring that has a strange anthropomorphic flavour: '. . . Damp, rust, dust and the fact that the spring always becomes weaker, because it is an inanimate thing, that is subjected to great stress. For living things are revived by food.

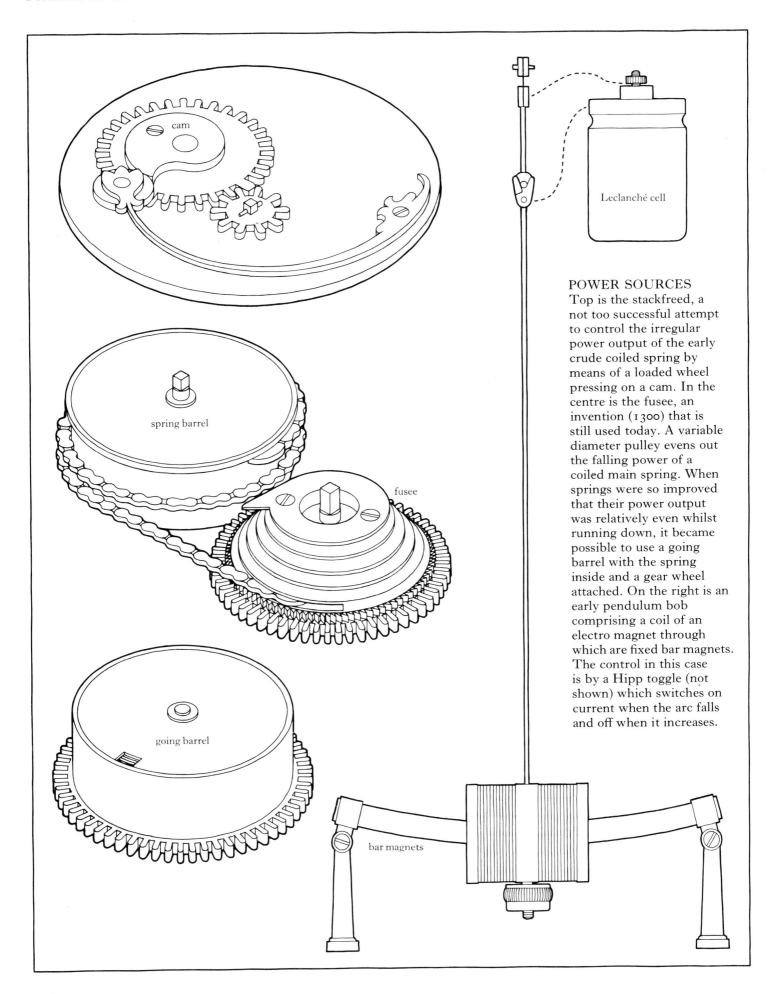

cam

Leclanché cell

spring barrel

fusee

POWER SOURCES

Top is the stackfreed, a
not too successful attempt
to control the irregular
power output of the early
crude coiled spring by
means of a loaded wheel
pressing on a cam. In the
centre is the fusee, an
invention (1300) that is
still used today. A variable
diameter pulley evens out
the falling power of a
coiled main spring. When
springs were so improved
that their power output
was relatively even whilst
running down, it became
possible to use a going
barrel with the spring
inside and a gear wheel
attached. On the right is an
early pendulum bob
comprising a coil of an
electro magnet through
which are fixed bar magnets.
The control in this case
is by a Hipp toggle (not
shown) which switches on
current when the arc falls
and off when it increases.

going barrel

bar magnets

Therefore all clocks, by lapse of time, become slower, never faster.'

The fusee for clocks was probably developed from a similar device used in war machines. Winding a huge crossbow to throw a massive arrow or a catapult to throw a boulder involved a similar problem. The spring was a beam that was bent, or it comprised skeins of sinews, or horsehair and human hair bought from poor women or donated during sieges. The skeins were twisted to act as springs for ballista that could throw a 25–30 kilogram (60–70 lb.) stone as far as 180 metres (600 feet) at the time of the Siege of Tyre in 332 BC. Men turned a windlass to wind the skeins, which required more and more strength as winding proceeded. The fusee, it is believed, was invented to take the strain off men's arms so they could wind the catapult more tightly.

There are illustrations of spring-driven clocks in a sixteenth-century notebook discovered by Professor E. Zinner in the Augsburg City Library some years ago, although most of the clocks described are weight-driven. The notebook was compiled by Brother Paul the German, known at the time as Paulus Almanus, during the ten years that he spent in Rome. During 1970–1, the Almanus manuscript was translated and interpreted by John Leopold of the Groninger Museum, in the Netherlands, and it shows that domestic iron chamber clocks were available in some variety in the fifteenth century.

Brother Paul, who lived in Rome from 1475 to about 1484, made himself an income by setting up as a clock repairer and supplier. His notes, which he made as personal memoranda of all the mechanisms he came across, include 30 descriptions of chamber clocks of all kinds and origins. Unfortunately, few examples of the clocks have survived.

Most of the clocks have 24-hour dials, only eight showing 12 hours. The 24-hour system of marking the hours was then current in Italy. Clocks ranged from simple alarms that were weight-driven and shook a bell at the appointed hour, to complex ones that played music and showed the date and the phases of the moon. They came from several countries, although Brother Paul did not give many makers' names. It seems almost certain that the simpler clocks were made locally and the more complicated ones were imported.

Eight were spring-driven, and all of these must have come from Burgundy or Flanders,

the main centres of the time. A weight-driven musical clock with alarm and 12-hour dial belonging to the Cardinal of Naples certainly came from Flanders. It had been converted to show time also on a 24-hour dial. Imported 12-hour clocks were sometimes converted to show Italian hours, and one described also had the striking converted to sound up to 24 strokes.

All the clocks described by Brother Paul had frames with plates top and bottom held together by vertical pillars and straps. Weight-driven and spring-driven clocks were of the same basic construction. Such iron clocks were to be found all over Europe. (One even turned up in London in 1975 behind the dial of an otherwise ordinary grandfather clock. It had been made by the Swiss family of clockmakers named Liechti.)

Above : A drum timepiece (i.e. not striking) to which a separate alarm mechanism has been attached. The single hour hand trips the pre-set lever below the alarm. The central hand on the timepiece shows the month. It is spring-driven and dated about 1600. A spiral spring was added to the circular balance some time after 1670.

Above: Movements of early clocks were decorative, like this table clock by David Bouquet, who was admitted to the Worshipful Company of Clockmakers in London in 1632. It is spring-driven and the count wheel, which controls the striking, can be seen at the top.

Right: A French miniature tower or belfry clock, 19cm (7½ ins) high, made about 1560. Inside the hexagonal case is a two-tier movement with a bell under the pierced dome at the top. The figure of Diana is engraved under the one-hand dial and there are other gods on the sides. The name 'Reiss, Prag' has been engraved on the dial at a later date.

THE CLOCK TRADE

In the sixteenth century the main market for mechanical clocks and watches was drawn from the new middle classes, comprising the craftsmen themselves, whose numbers were growing rapidly, the merchants who sold their products, and professionals such as apothecaries, doctors and lawyers.

At this time the pioneering clockmakers of the Duchy of Burgundy and of Flanders lost their lead to two towns on the *alte romantische Strasse* in what is now south Germany, Nuremberg and Augsburg. In the early sixteenth century Nuremberg became the major manufacturing centre for clocks and watches, then Augsburg gradually took over in the second half of the century. The clockmakers of Augsburg produced and exported an incredible variety of clocks, some of them with complex astronomical dials. The astronomical clocks were square or drum-shaped table clocks with the dials on top and plated movements. Often the sides had up to eight subsidiary dials.

Some Augsburg makers produced spring-driven globes and spheres to demonstrate astronomical features as well as to indicate the time. Others made monstrance clocks — shaped like a hand mirror on a stand with elaborate decoration. The tabernacle clock was a favourite form. This is a monumental clock in gilded metal up to 90 centimetres (3 feet) high, with a tower on top enclosing the bell and a series of canopies. Often there were two dials, one for the hours and the other for quarter-hours, or a 12-hour and a 24-hour dial.

Showing an early flair for developing their export markets, Augsburg makers also produced a variety of novelty clocks with models of animals holding the dials. Some were animated, the eyes of the figure swinging with the movement of the balance in the clock. More elaborate versions included an animated maid milking a cow from which milk, previously inserted, squirted into a bucket. Another was a moving ship.

In the meantime, the Flemish and Burgundian clockmakers developed table clocks that were much smaller than those of their rivals. Clocks were also being made at an early period in Italy, but perhaps mainly for domestic consumption.

In the later sixteenth century, the Augsburg trade was already so well organized that members of different crafts supplied various parts of a clock. The bronze-founder and the brass-beater co-operated on some

Right : A monstrance clock, made about 1690 in Lindau by Gottfreid Geiger, 55cm (22½ ins) high. A true monstrance is used by the Roman Catholic church as a receptacle for the consecrated host, being a glass or crystal case in an elaborate metal frame. The pendulum was added later.

cases, and the goldsmith and engraver on others. The clockmaker probably relied on a specialist for his springs.

It is known that at least one workshop supplied special ornamental strip for making cases in the 1560s, and that the strip was probably exported. Nine sixteenth-century table clocks of drum or square form with sides of the identical ornamental strip are known at present. The elaborate design in relief depicts Orpheus and Eurydice with many animals. The movements of these Orpheus clocks vary considerably; they came from several workshops. Dials are elaborate, and it is likely that the dial on top set the diameter of the round case or the length of side of the square one. The casemaker then took the strip he had bought from the strip supplier and cut it to length. Where the ends were joined, he tried to make the pattern match as unobtrusively as possible. If a length of strip was too small, he would insert a strip from another patterned length.

One of the Orpheus clocks has a brass instead of an iron movement and has a table for calculating Italian hours engraved on it. There are other indications of Italian origin. The ornamental strip may have been exported to an Italian maker, or the clock case may have been made in southern Germany for export to Italy.

Henry VIII of England gave his fifth

wife, Catherine Howard, three tablets of gold 'wherein is a clocke', and he also presented her with one of the earliest watches mounted in a gold pomander, which had certainly been imported, probably from Nuremberg, or from Blois in France.

Export of jewellery from Augsburg, Nuremberg and several other centres was already considerable in the sixteenth century, and Queen Elizabeth I appointed a 'High Germaine', Master Spilman, as her chief jeweller. The making of timekeepers involved similar craftsmen, and naturally the same lines of commerce.

Elizabeth I had a complete section of her inventory of 1572 devoted to watches, but they were still categorized as jewels. Some were pieces of jewellery, such as pendants set with watches, and others were 'form

Right : In this automata clock, the lion's eyes are moved to and fro by the balance (1670).

Left : Dial of a spring-driven Augsburg table clock (1736) by Johann Beitelrock, which is just over 41cm (16 ins) high. In the centre is an astrolabe dial indicating the visible part of the sky at different times of the year.

Below : Jacques de la Garde made this pomander watch in 1551.

Below : An Orpheus clock, named from the pattern on the sides of the case, showing Orpheus charming the animals with his flute. On the top is an alarm attachment, set off by a release lever, not the cruder form of trigger operated by the hour hand. Nine Orpheus clocks are known to exist.

watches', in the shape of crystal crosses, flowers, *memento mori* (death's heads), as well as the early pomander watches later misnamed 'Nuremberg eggs'.

Some of these articles were probably gifts, but many were bought from travelling merchants, who sometimes had competition from smuggled goods, as they do today. Of the many merchants who visited European countries and Britain, 'a great number of rascals and pedlars and juellers' entered England in 1518, following the French ambassadors and gentlemen of the court. With them they brought 'divers merchandise uncustomed, all under the colour of the Trussery of the Ambassadours'.

One method of encouraging orders was to send a drawing, the prototype of today's catalogue. In Henry VIII's time the great

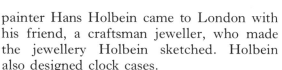

Left : When the
portable clock became
the watch, it often
became a piece of
jewellery also,
sometimes a macabre
piece like this
memento mori, carried
on the belt or on a
necklace. The silver
skull with watch inside
is not much more than
3cm high and was
made in 1658 by
Abraham Caillatte.

Below : An example of
a personal portable
timekeeper in the form
of a silver cross, made
about 1580 by Melchior
Zinng, Augsburg.

painter Hans Holbein came to London with
his friend, a craftsman jeweller, who made
the jewellery Holbein sketched. Holbein
also designed clock cases.

Designs were often circulated. In 1546 a
London merchant received from his Antwerp
agent a coloured drawing of a pendant set
with a large table-cut diamond with the
suggestion that he show the 'pattern' to the
King, as 'the time is unmeet to pester the
King with jewels. . .'.

It seems that design pirating was rife, too.
In 1562 Hornick of Nuremberg engraved
designs for jewels and ornamental watches.
These were almost identical with designs
in a list of jewels stolen from two Antwerp
goldsmiths in London the previous year.

Trade often had political overtones.
Elizabeth I was concerned with the growing
power of Spain and tried to enlist the help
of the Sultan of Turkey by sending him
presents, but she insisted that the Levant
Company, the English concern set up to
trade with the East, paid for them. For 200
years goods from the East had come through
the Mediterranean to Venice and been
shipped from there to England on 'Flanders
galleys', but Turkish sea-power had grown
so much that the eastern Mediterranean was
seething with corsairs who virtually cut off
the trade by plunder. The English merchants
wanted to negotiate safe passage with the
Sultan, but the French and Venetians were
installed in Constantinople and tried every
means of preventing it.

It was planned to send a secret and
dramatic present to the Sultan from the
Queen, and what more exciting gift than a
very special clock? It was to be combined

with an organ to play tunes before striking, have various animated toys and be decorated with gold, silver and precious stones. Thomas Dallam (*c.* 1570–*c.* 1630), a man who had come to London from Lancashire at about the same time Shakespeare came from Stratford, built the organ and was also given the task of taking it to Turkey and erecting it. Fortunately for us, Dallam wrote a diary recording his adventures from the day he set out on 9 February 1599. Even before reaching Constantinople, the little sailing ship encountered pirates in the English Channel, the captain was kidnapped

in Algiers and rescued by Dallam, and Dallam had to escape from Turkish slave-traders in Rhodes.

The monumental clock showed the time of day, the course of the sun and the moon, the moon's phases and age and the reigning planets. An automaton in the form of an armed man in one of the towers struck a bell loudly at the hours. A cock made of metal flapped its wings and crowed at the hours, and a tune was played on 16 bells. Eight figures made obeisance to a model of Queen Elizabeth set with diamonds, emeralds and rubies, and two trumpeters played a

Above : The painter Hans Holbein the Younger also designed jewellery and clock cases in London after he arrived in 1526. In this picture, entitled The Ambassadors, *he painted the exact time and date – 10.30 a.m. on 11 April 1533 – on the dials on the rug.*

Above: A gilded copper drum clock with hand and bell on top, made in England about 1581. It is thought to have belonged to Queen Elizabeth I.

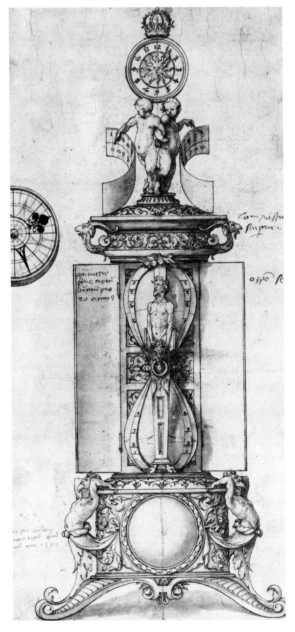

Left: One of Hans Holbein the Younger's designs for a clock. It was made after his death in 1543 and presented to Henry VIII.

fanfare. In addition, the eyes of a large head moved and an angel turned over an hourglass.

The organ was played automatically four times a day, and also had a keyboard for an organist to play it at any time. It, too, had its automata. Dallam wrote, 'in the tope of the orgon, being 16 foute hie, did stande a holly bushe full of black birds and thrushis, which at the end of the musick did singe and shake theire wynges'.

When the organ clock was unpacked in Constantinople, there was panic. Everything was mouldy, joints had opened and some woodwork split, the paintwork was cracked and blistered, and all the organ pipes were bruised and dented. The ambassador was beside himself with rage and fear and shouted at Dallam that the lot was 'not worthe 2d'. Imagine Dallam's despair. It had taken two and a half years to get permission for the present to be made, and another two years to make it. Now, after an incredible journey, it was smashed.

Dallam, however, was a tough craftsman. His reputation was at stake and he would not accept responsibility for the bad packing. The ambassador's taunt that the organ clock was not worth twopence made him lose his temper. 'My answeare unto our Imbassader at this time I will omitt', he wrote in his diary. Further unpacking revealed, fortunately, that the mechanisms were relatively undamaged.

Eventually, Dallam restored the organ and set it up in the Sultan's palace. His go-between was the Chief White Eunuch, and when the presentation was eventually made, no one from the party making the presentation was allowed in, so the Chief White Eunuch had to explain and operate the mechanisms. The Sultan sat immediately in front of the organ clock and was flanked by 100 imperial pages. On one side of the pages were 100 deaf mutes, many of them falconers with their falcons on their wrists, and on the other 100 dwarfs with scimitars. All were dressed in gold cloth.

This is known because the organ clock so pleased the Sultan that he called for Dallam, who later wrote that he was terrified he might be killed because he was told to sit and play the organ with his back to the Sultan; moreover, he would touch the Sultan's knee with his breeches. Both crimes were punishable by death. So was disobeying the order. He hesitated and the Sultan spoke again, so the Chief White Eunuch thrust Dallam forward. As he sat at the organ, the Sultan rose to have his chair moved and he knocked into Dallam, who thought he was 'drawinge his sorde to cut off my head'.

But all was well, and Dallam played popular Elizabethan tunes and 'suche thinge as I coulde untill the cloke strouke, and then boued my heade as low as I coulde, and wente from him with my backe towardes him'. This caused him more trepidation, but the Sultan was so pleased he gave him some gold coins. It was not until he left the room that Dallam suddenly remembered the ambassador, who had been unable to present his credentials and had waited elsewhere in a fever of anxiety for two hours.

Dallam's next fear was that the Sultan would keep him in Turkey, to maintain the organ clock and to move it to different places, but he got away eventually on the same ship, the *Hector*. The present of the clock heralded a later extensive trade in clocks and watches with Turkey.

THE CLOCKMAKERS

In Augsburg, clockmaking was associated with lockmaking, but was considered an inferior occupation because locksmiths, gunmakers and certain other craftsmen were

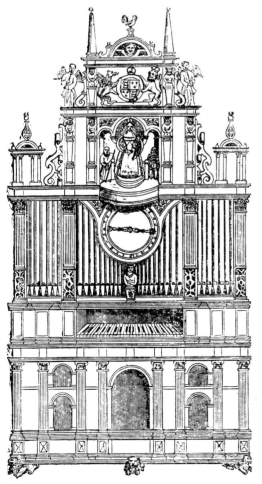

allowed to make clocks, but clockmakers were not allowed to make locks and guns.

Nuremberg's clocks were also made by locksmiths until a clockmakers' guild was formed in 1565. The town council set a test for those wishing to join the guild. They had to make a clock and a watch within a year. The clock had to be 15 centimetres (6 inches) high and must strike the hours and the quarters. The dial had to show the 24 hours of sunshine and moonshine on one side, and also indicate the quarters. The other side had to show a calendar and the planets with the length of the day. The watch had to be 'such as one wears hanging at the neck', and to incorporate an alarm. These entry tests, which were later made even more severe, must surely have eliminated most locksmiths. Nevertheless, locksmiths remained entitled to make clocks until as late as 1629, although clockmakers were still banned from making locks.

The main competitor to Nuremberg for watches was Blois in France, where the industry was established not much later. There were turret or public clockmakers as well as locksmiths making clocks in Blois before 1500. The earliest watchmaker recorded is Julien Coudrey (d. 1530), maker to King Louis XII. Coudrey produced two *orloges* for the King, one probably a watch set in the hilt of a sword. From about 1515 to 1610, the number of workshops grew from about 5 to 63. The English diarist John Evelyn visited the town in 1644, and wrote about the fine air and courteous inhabitants, who were 'so ingenious, that, for goldsmiths' work and watches, no place in France affords the like'.

Paris was another early centre, its guild being formed in 1544, 53 years before that of Blois, despite the early royal patronage of Blois clockmakers. There seems to have been an exceptionally strong element of protection in the Paris guild rules. An apprentice had to pay such a large sum of money on becoming a master that it gave an unfair advantage to the sons of masters. Nevertheless, after the decline and elimination of Blois and other towns – Rouen, Lyon, Angoulême, Grenoble and Sedan – as watchmaking centres, Paris still continued as an important clockmaking centre.

Because of the calculations that had to be made, a clockmaker or a watchmaker needed a higher degree of literacy than was required by any other sort of craftsman, such as a blacksmith or whitesmith, a locksmith or a jeweller. Masters were required to teach

Left : A Victorian engraver's idea of what Thomas Dallam's organ clock for the Sultan looked like. It is, however, more a reflection of Victorian than Elizabethan ideas.

Left : A sixteenth-century engraving of a clockmaker's workshop in Flanders, attributed to Stradanus. There is a forge in the background, but no bench for the making of parts by sawing and filing is shown. Weight-driven clocks and watches hang from the walls. A watchman's clock and a verge and foliot are propped up against the bench.

Left : A small silver watch with one hand, made about 1600 in Blois by Pasquier Peiras. It measures $3 \times 5cm$ ($1\frac{1}{4} \times 2$ ins).

apprentices to read and write in the first year of their apprenticeships. Apprentices had to be able not only to draw but also to count and do basic mathematical calculations. Not a few studied books of astrology and were able to perform complex calculations. Because of their literacy, a higher proportion of clockmakers and watchmakers followed reformist movements and therefore they suffered more from religious persecution.

The emigration of a few key clockmakers and watchmakers at that time could alter the balance of power between nations. Although clockmakers and watchmakers comprised only a very small part of the total number of refugees, the knowledge and skills that they took with them had a disproportionate effect on the manufacturing industries of the countries where they sought refuge. This was because so many associated crafts were affected by the capacity to make clocks and instruments. Even astronomers and thinkers were influenced, when considering where they would be able to work best, by a country's capacity to produce accurate instruments.

The Frenchman Jean Cauvin, or Calvin, who spread the Protestant Reformation in France and Switzerland, managed to get a

Presbyterian government established in Geneva. Geneva had fallen on bad times. Its merchants had drifted away and industry was stagnant, so Protestant craftsmen were welcomed and given special privileges. After 1550, clockmakers began to arrive in Geneva as refugees from France, plus a few immigrants from Germany, the Low Countries and Italy. Similarly, Henry VIII and Elizabeth I allowed many immigrant craftsmen to settle in London.

Augsburg and Nuremberg began their long decline during the Thirty Years' War from 1618 to 1648. French centres benefited, but not for long, because political persecution of Protestants was renewed when the Edict of Nantes, which gave them some protection, was revoked in 1685. A new wave of refugees left France to seek asylum in Geneva and London.

The balance of power was now tilted in favour of England and Switzerland. The effect was at first greatest in England, where the growth of clockmaking and watchmaking coincided with, indeed was part of, the scientific revolution, and there was considerable co-operation between makers and the new scientists of the Royal Society. A stream of fundamental inventions resulted, and before 1700 London became the clockmaking and watchmaking centre of the world.

The refugees with foreign names brought fresh ideas, prosperity and also means of export with them, but they were not welcomed universally. Some Calvinistic Genevans disapproved of the fact that refugees were attracted to the city more because they could practise their craft there than by the Gospel. By 1600 about 25 master clockmakers worked in Geneva, and 80 years later there were over 100, who were responsible for producing about 5000 watches and clocks a year.

In England, a Bavarian, Nicholas Cratzer, became clockmaker to Henry VIII. Elizabeth I chose the Frenchman Nicholas Urseau. The first indigenous Royal clockmaker was a Scot, David Ramsay, appointed to James VI of Scotland, who became James I of England on Elizabeth's death. It was recorded that the English 'prefer strangers as well as phisitians as other like professors then their owne countrymen as more learned and skillful than they are, which makes the English also so much travayle in forrayne parts. . . ' . That did not prevent English craftsmen from resenting the strangers.

Sixteen clockmakers petitioned the King in 1622 to rule that foreigners as well as Englishmen had to start under English masters. They were 'much agreed both in theire estates credittes and trading through the multiplicitie of Forreiners usinge their profession in London . . . whereby their Arte is not onelie by the badd workmanshipp of Straingers disgraced, but they disinhabled to make sale of theire comodities at such rates as they maie reasonablie live by'.

It had no result; neither did a petition of 1627, made through the Worshipful Company of Blacksmiths. In 1632 the clockmakers succeeded in a petition to form their own

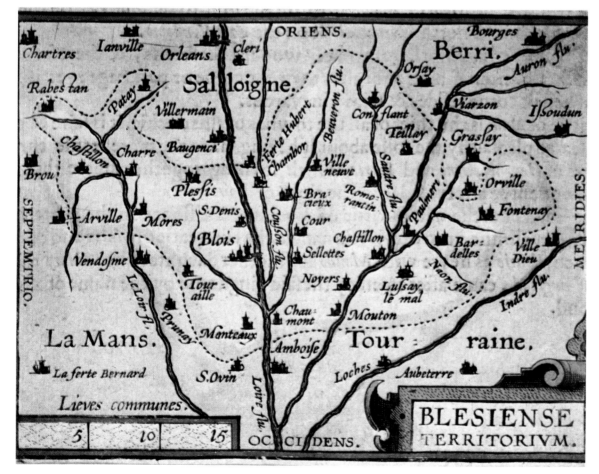

Left : A map of Blois in France and the surrounding territory from the 1590 edition of Theatrum *by Abraham Ortelius. At the time, there were about 28 watchmakers' workshops in the town. A Guild was formed seven years later.*

MEASURING TIME

During the first centuries of the domestic clock, methods of measuring time varied in different countries. Brother Paul handled clocks with both 12-hour and 24-hour dials. In Italy then time was measured by what are now called Italian hours, starting at sunset, or about half an hour after dusk, and running through 24 hours until the next sunset, so that the morning hours began at about 12. A clock indicating only from 1 to 12 was confusing and was usually converted. The famous clock at Milan, dated 1335, which was probably the first to strike actual hours instead of one blow at the hour, struck from 1 to 24.

Up to the fourteenth century, all countries divided daylight into a certain number of temporal hours (usually 12) and night into the same number. The change to regarding day and night as a single period of 24 hours seems to have occurred in each European country at about the same time as it adopted the mechanical clock. The 24 hours were not always counted as 1 to 24 as in Italy, but in some countries, including England, in two periods of 1 to 12 which were shown on one dial.

More important than the way of counting was the point of time from which hours were counted. To measure time accurately from dawn or dusk was difficult. With the double-XII system, it became usual to reckon from noon. Noon was relatively simple to establish by the position of the sun or a shadow it cast.

It is known that double-XII hours were reckoned from noon in England in the late fifteenth century, because the poet Chaucer,

guild and statutes, and with this came the right to control foreigners, who had to become members of the guild in London. The first court of the Master, Wardens and Fellowship of the Art or Mystery of Clockmaking of the City of London was held in 1632. This became the Worshipful Company of Clockmakers, through which London makers controlled the craft for more than a century by setting standards and rules for apprentices and by having the power to enter premises to break up badly made clocks, watches and other timekeepers.

The Clockmakers Company is still active, but in a different way, and has a great many clockmakers, watchmakers and collectors among its Freemen and Liverymen. It possesses an excellent collection of clocks and watches as well as a library.

Bottom right : An Italian spring-driven clock of about 1730 which has a I to VI dial, presumably so that power was saved in striking.

Below : Nicholas Cratzer, 'deviser of the King's horologies and astronomer' to Henry VIII, painted in London in 1528 by Holbein, who helped him to design cases. He is believed to have lived in England for 30 years without learning to speak English.

in a treatise on the astrolabe directed to a child of ten, referred to 'xi of the clokke by-form the houre of noon til on of the clok next folwyng'.

Another advantage of the double-XII system was that less power, and therefore less weight fall, was needed for striking. The Italians had this problem, as they persisted with their 1 to 24 system for centuries, although the rest of Europe had changed to the double-XII at the beginning of the fifteenth century and had begun the change to a single-XII dial 150 years later.

Referring to 1 to 12 striking as French, or sometimes *oltramontane* (beyond the mountains), the Italians attempted to solve the problem of providing power for striking by inventing their own system using a 1 to 6 clock dial and striking to suit it, so that the clock's hour hand turned four times during the course of a day. It was up to the owner to decide whether 3 o'clock, say, was that time at night, 9 o'clock in the morning, 3 o'clock in the afternoon, or 9 o'clock in the evening. In the eighteenth century, two English makers, Joseph Knibb and Daniel Quare, introduced what was referred to as 'Roman striking' for the same purpose of reducing the number of hammer blows and thus the power storage of the weight, but this was quite different from the early Italian method. They fitted a bell with a high note representing the Roman I and another bell with a low note representing the number V. A low note followed by two higher ones was therefore VII. Two blows on the lower bell meant X, so a high note followed by two low ones indicated IX. The system was

valuable for clocks running a month at a winding because the number of hammer blows was less than half the 5000 or so normally provided for.

The odd fact is that, although it seems that one purpose of the Italian clocks was to save power, they had double striking, so the power consumed was the same as that required by a 12-hour clock. The hour struck was repeated after an interval of two or three minutes. Repetition of the hours – *la ribotta* – is still commonly heard from turret clocks in Italy as well as other places in Europe.

The reserve of power of the portable clock was of course even more limited than that of the weight clock. It was probably this fact more than any other influence that hastened the change to the universal I to XII striking and dial.

65

IV
EUROPEAN MECHANICAL CLOCKS

The Gothic iron clock for the household changed little until the later sixteenth century, when parts of Europe were on the brink of a new age that brought a new scientific spirit of enquiry and an explosion of knowledge and activity.

THE LANTERN CLOCK

English clockmakers, and continental clockmakers working in London, had, shortly before 1600, developed their own style of weight-driven clock, now known as the lantern clock. Although the clock is lantern-shaped, and the first ones were made of iron, by about 1600 brass became standard, so the name lantern may be a corruption of *latten*, meaning brass. They were referred to in old inventories as 'brass clocks'. A lantern clock was hung on the wall and went for about 12 hours before the weights hit the floor. A single hand pointed to the hour on a I to XII dial and the hours were struck on a bell at the top. Frequently the clock had an alarm, set by a small dial in the centre of the main dial that was turned so that the pointed tail of the hour hand pointed to the hour at which the alarm was needed. The alarm disc rotated with the hour hand.

The escapement was a verge and crown wheel always combined with a one-spoked wheel balance. The movement was enclosed by brass panels to keep out the dust, but the dial and bell on the top were exposed. In front of the bell there was usually a characteristic fret, most often depicting two dolphins but sometimes with a heraldic motif. On the back was a stirrup (a half-hoop to hang on a large hook) and two pointed spurs to dig into the wall and prevent rocking

while the clock was wound. Each weight cord ran round a grooved pulley in the clock and had a heavy driving weight at one end and a smaller weight at the other to keep the cord in the pulley groove.

The lantern clock movement had four posts at the corners. In this it resembled its predecessor, the domestic iron clock, which was based on the big medieval cathedral

Left: A musical bracket clock of the George III period made for the Turkish market.

Below: Two lantern clocks (1653), that on the left has an alarm as well as striking.

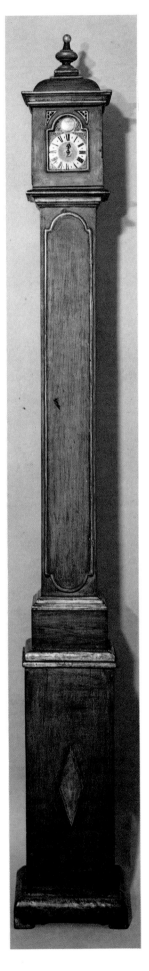

Right : When the long
wooden case was
introduced about 1660,
many lantern clocks
were enclosed in long
cases instead of being
hung on the wall, like
this one by Asselin,
London, c. 1700.

clock, but the posts were turned instead of being flat or like buttresses; they were more like the bedposts of the later forms of turret clock. A major difference was that there were plates at top and bottom to unite the posts, instead of bars, as in the earlier iron clocks. Plate frames had already been in use for some years in Europe.

Brass had at least two advantages. It could be cast, a much quicker process than being heated to red in a forge and hammered into shape, as was necessary with forged iron. Also, when iron pivots worked in brass holes, the friction was lower and the wearing properties much better than when iron worked in iron.

There was no sophistication about the timekeeping, however. The lantern clock seems to have been a simplified version of the Gothic iron clock made to reduce costs and expand the market for clocks. The wheel balance could not be adjusted like the foliot, so the rate of the clock was altered by varying the driving weight. Some weights had cups in the tops for adding lead shot.

THE PENDULUM CLOCK

The lantern clock was made in more or less standardized form until about 1660, when there was a change in clockmaking in England that had repercussions throughout Europe. Yet the man responsible for the change was not English or even working in England: he was the celebrated Dutch astronomer and physicist Christiaan Huygens (1629–93), who invented the pendulum clock in about 1657.

Augsburg, at the beginning of a slow decline, was still very active. (Later it was to concentrate on *Telleruhren*, which were clocks made like shields of beaten metal, popular as wedding presents in Germany.) The clockmakers of Flanders and Burgundy still could not compete with the industrious southern Germans. However, despite having no long history of clockmaking, England surged rapidly ahead. This advance was largely owing to the ideas and experiments of members of the new Royal Society. Moreover, in the seventeenth century the British were not only inventive themselves, they also exploited the inventions of others. Only a few months after Huygens announced his pendulum clock, an English clockmaker, Ahasuerus Fromanteel (1607–1693), obtained a right to make them. He advertised them in the *Mercurious Politicus* and the *Commonwealth Mercury*.

The clocks were described as 'keeping an equaller time than any now made without this Regulator (examined and proved before his Highness the Lord Protector by such Doctors whose knowledge and learning is without exception) . . . and may be made to go for a week, a month, or a year, with one winding up. . . '. The Regulator was the pendulum. The precision introduced by the pendulum coincided with the rigid religious disciplines introduced by the Puritans, and Oliver Cromwell, the Lord Protector, was sufficiently interested in it to approve it. Once more timekeeping had found a sponsor in a religious disciplinarian.

The invention was taken up by London clockmakers in an incredibly short time, considering the fact that in those days ideas could take years to spread to a neighbouring town. In about 1660 London makers invented the longcase (grandfather) clock, with a short pendulum of about 25 centimetres (10 inches). It stood about 1.8 metres (6 feet) high in a good quality ebonized or walnut case which enclosed the weights. When the long pendulum, also called the Royal Pendulum, because it dominated the clock, was introduced about 1670, the longcase clock became taller, as well as wider.

Above right : The movement of one of the first longcase clocks, made in London c. 1660 by Ahasuerus Fromanteel. It has plates front and back, unlike the lantern movement with corner posts, but retains the verge escapement with short bob pendulum.

Right : Probably the earliest clock with an anchor escapement and long pendulum. It was made in 1671 in London by William Clement for King's College, Cambridge, and is now in the Science Museum. Bolt and shutter maintaining power keeps it going whilst being wound.

The movement of the longcase clock was like that of a lantern clock turned sideways, so that the plates were at front and back and the pillars horizontal. This arrangement had the advantage that the arbors of the wheels could be planted in any position, not being confined to rows in a bar, and that a key could be used for winding the weights through holes in the dial.

The humbler lantern clock did not disappear. It too went inside a long case. London makers concentrated on longcase clocks that ran for a week or more at a winding. Local makers used lantern clocks in long cases that had pull-up winding and ran for about 30 hours. Later the same kinds of movement with plates front and back were used for 30-hour (daily wound), 8-day (weekly wound), one-month, three-month, and one-year clocks. After about 1750 the longcase clocks were made all over the British Isles, but more in Yorkshire than anywhere else, and London, Birmingham and Newcastle had embryo factories turning out movements for the clockmakers.

The long pendulum was about 1 metre (3¼ feet) long, and it swung from one side to the other in one second – in clockmakers' language, it had a beat of one second. Its use was made possible by the invention in about 1670 of the anchor escapement, which has been attributed to the clockmaker William Clement, (*fl. c.* 1671–99), of London. He made a turret clock with an anchor escapement in 1671 for King's College, in Cambridge, where it remained until 1817, when it was moved to St Giles Church, and eventually to the Science Museum, London.

Alternatively, the originator may have been Joseph Knibb (d. *c.* 1711). A few years ago, a wrought-iron turret clock with a form of anchor escapement was found in the church at Burley, Leicestershire. The top bar frame bears the inscription 'Joseph Knibb London 1678'. This date is of course later than that of the King's College clock, but Knibb was known to have been experimenting with such escapements for some time. There is a reasonable possibility that Dr Robert Hooke had a hand in the design, as claimed by William Dereham, author of *The Artificial Clockmaker*, in 1696. Dereham wrote that Hooke denied that Clement was the inventor, but no other evidence has been found.

It is said that Clement was once an anchor smith, which gave him the idea of the anchor shape, but this is in doubt, too. The anchor escapement has pallets where the

points of a real anchor would be. The escape wheel is, flat, unlike the crown wheel, but the teeth are of the same shape: in fact, it is like a flattened-out crown wheel. The anchor is linked by a loose connection called a crutch (invented by Huygens) to the pendulum, which is suspended separately.

As the pendulum swings, the anchor, which spans several teeth of the escape wheel, usually seven, allows a tooth on one side to escape, then, a second later, a tooth on the other side. As each tooth is released, it gives a push to the anchor to impulse the pendulum and keep it swinging. English makers used an escape wheel of 30 teeth so that it turned once a minute in 60 jumps and a seconds hand could be attached to it.

English makers of spring-driven clocks also enthusiastically adopted the pendulum,

Below : A bracket clock by Thomas Tompion, with a mock pendulum, four subsidiary dials, and winding key. It has grande sonnerie striking, which means it strikes the hours and quarters at every quarter of an hour.

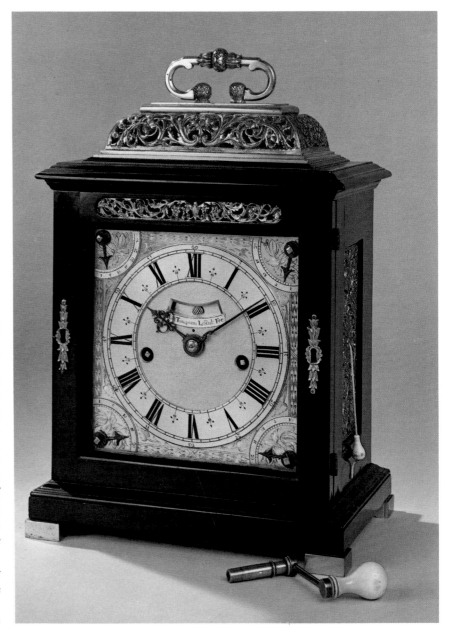

TRAINS

Below is a weight-driven English striking clock.
The going train (right) is controlled by an anchor
escapement and pendulum (not shown). It turns the
12 to 1 geared motion work (bottom right) operating
the hands. The smaller wheel of this has a pin to
release the striking train (see rack striking page 44).
The striking train on the left is controlled by a
count wheel or snail and rack. Its second wheel
carries a series of pins to operate the hammer and
the speed of striking is controlled by the fly (air
governor) at the top. Right is the gear train of a
fusee pocket watch with an English lever escapement
and plain balance. The centre wheel turns motion
work, on the other side of the plate, to which the
hands are attached.

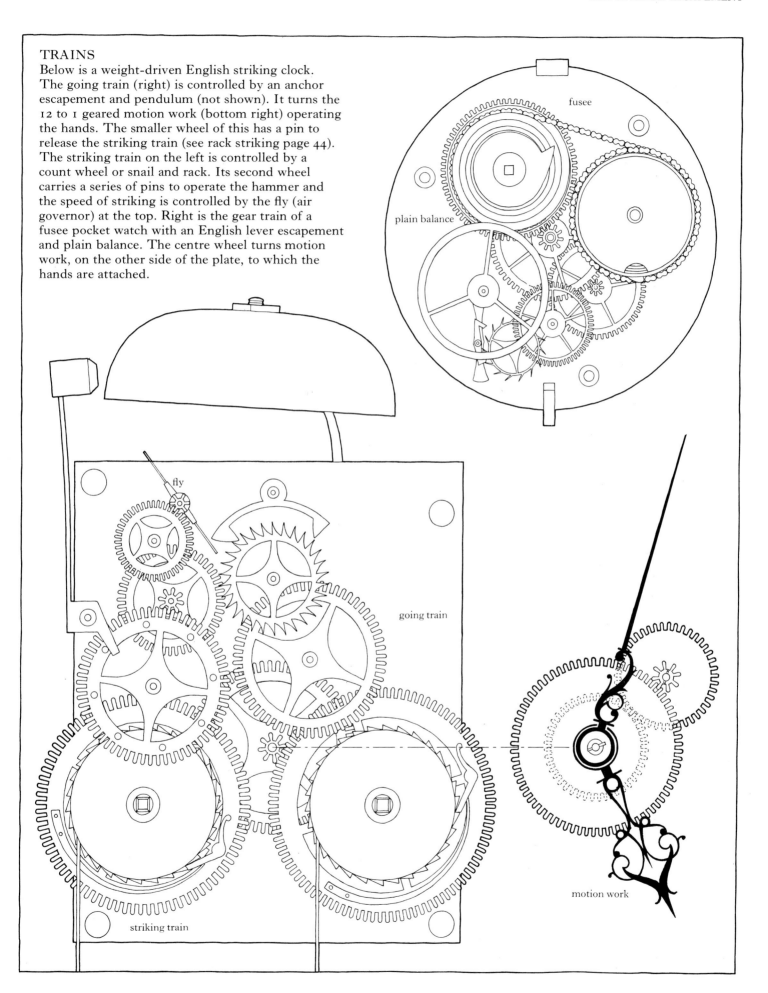

fusee

plain balance

fly

going train

striking train

motion work

but in the much less accurate form illustrated in the sketches of Leonardo da Vinci (although they could not have seen these), where the pendulum rod was attached directly to a horizontal verge working in a horizontal crown wheel. It was not a very satisfactory arrangement, but continued for more than a century.

The application of a pendulum directly to a verge escapement made conversion of balance wheel and foliot clocks relatively simple, although in English clocks the crown wheel usually had to be changed from a vertical to a horizontal position. The Dutch had a simpler solution, which left the crown wheel where it was.

The pendulum swept through all clock-making countries. Many thousands of old clocks were converted, and those under

construction were redesigned for pendulum control. English lantern clocks were converted to long rather than to short pendulum, to provide better timekeeping at the expense of incongruous appearance. Many old Augsburg clocks were converted by hanging a short, wide-swinging pendulum in front of the dial, which looked almost as strange.

At last here was a clock that could be made relatively easily and, in its most accurate versions, with weight drive and long pendulum, would keep time to seconds a day. The astronomers were the main beneficiaries. It also helped the map-makers, but it did nothing for the sea navigators. A pendulum would not work at all accurately at sea, as Huygens discovered.

The longcase clock, which, as it turned out, remained accurate for centuries, was copied everywhere. The wooden, short-pendulum bracket clock also set a new style, most domestic clocks before it having been metal-cased.

CLOCK CASES

The first English pendulum clock cases were classical, with hoods based on the Parthenon. Sometimes they had barley-twist Jacobean pillars instead of Ionic ones, and the cases were often in veneered ebony. Styles developed as new woods were introduced. There was a period of complex marquetry and also a fashion for japanned cases, but, in general, English cases never reached the florid extravagances of some other countries. In the later stages, mahogany, and oak for cheaper clocks, became most common for long cases.

English makers devoted themselves to quality and to perfecting the movement and its production. They were less interested in the cases than their continental rivals were, and very little is known of the casemakers' trade in England. Tentative steps were taken towards batch production, but the individual craftsman attitude dominated, a fact that was to put most English clock-making and watchmaking out of business in the twentieth century.

An exception to the generally conservative development was the specialization by a few London makers in complicated automata clocks, mainly for Eastern markets. These makers were following the lead of Thomas Dallam.

The most successful and ingenious of these makers was James Cox (d. 1788), whose workshop turned out musical as well as automata clocks. One of his masterpieces

Left : A bracket clock (c. 1760) by John Ellicott, London, in green lacquer. In the arch is a strike/silent control, and in the dial a mock pendulum to show that the clock is going, and a date aperture.

Below : A longcase chiming clock (c. 1740) by Isaac Nickals of Wells, Norfolk. There is a tidal dial around the moon dial in the arch, and a centre-seconds hand. The case is in gold and lacquer.

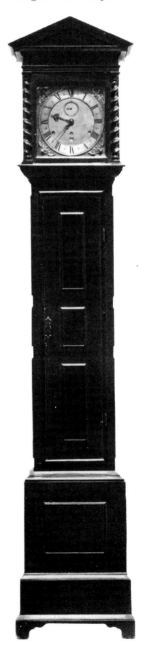

Far right : An automatically wound clock (c. 1760) by James Cox, London, who made many musical and automata clocks specially for export. The winding is performed by a Fortin mercury barometer in the case, which moves with changes in air pressure.

is the peacock clock, now in the Pavilion Salon of the Hermitage Museum in Leningrad. A huge peacock sits on top of the pruned trunk of an oak tree with some branches bearing leaves. The whole clock is about 2.5 metres (8 feet) high. A cage containing a silver owl hangs by a silver chain from a dead branch, and a squirrel sits under the cage. The surround is like a flower bed of gems represented by coloured foil and surrounded by green lacquered 'turf'. Six mushrooms grow in front of the bed. Roman figures for the hours and Arabic ones for the minutes appear on the cap of the largest mushroom, while seconds are recorded by circling grasshoppers. Other mushrooms, acorns and leaves lying on the ground conceal winding holes. There are also other snails, lizards, a frog and a snake.

Left : Floral marquetry decorates the case of this longcase clock by Charles Gretton, London, c. 1670–1695. The clock runs for a month at a winding and was in the famous Wetherfield collection sold in 1928.

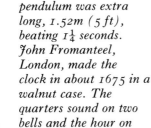

Right : One of the earliest longcase clocks with an anchor escapement. The pendulum was extra long, 1.52m (5 ft), beating 1¼ seconds. John Fromanteel, London, made the clock in about 1675 in a walnut case. The quarters sound on two bells and the hour on four simultaneously.

Bells sound the hours and the quarters, and at a certain time the owl begins to move, its cage turns, sounding bells which are attached to it, while the owl blinks its eyes, turns its head, and beats time to the music with a claw. Other birds join in, then, as soon as the owl's performance is finished, the peacock slowly raises and opens its great, shimmering tail, then lowers it and bows its head gracefully. (This was once done to music, which has since failed.) Lastly, a cock crows hoarsely.

The clock probably went to Russia with the Duchess of Kingston, who arrived in St Petersburg in 1777 with a load of art treasures from England. It seems to have been transported in pieces, which were then left in a palace storeroom for a considerable time, because various pieces were missing when a Russian mechanic, Ivan Petrovic Kulibin, was given the task of assembling it and setting it up.

Cox made most of his automata clocks for China. There are eleven of them still in the Winter Palace at Peking and seven in the Peiping Museum, most of them elaborate. He also made a monumental longcase clock that was automatically wound by a Fortin barometer made of glass and containing 68 kilogram (150 lb.) of mercury, suspended in the trunk of the case, and a clock that was wound by opening and shutting a door. The barometer clock is still in England, in the Victoria and Albert Museum.

The first pendulum clocks, made for Huygens by Salomon (Samuel) Coster (d. 1659), a clockmaker working in The Hague, were wall clocks in cases rather like picture frames, the framed glass front being a door. The Dutch continued to make such clocks, which they called *Haage clokjes*, and the French made some similar ones, too, but neither nation did so for long. The style really came into its own when it reappeared in the early nineteenth century in the form of the American shelf clock.

The Dutch in Friesland made clocks with movements similar to those of English lantern

Left : A French clock of the religieuse *type – the case is decorated with Boulle inlay work. The maker was Jacques Thuret, Paris, in about 1700.*

Right : The dial of a Dutch automata longcase clock, with a moving marine scene at the bottom, calendar sectors, and a moon dial. The arcaded minute ring is a feature of many Dutch clocks.

clocks, but with a cast fret, much larger than the English version and highly decorated, completely hiding the bell. The clock, called a *stoelklok* (stool clock), instead of being hung directly on the wall stood on a wall bracket that was also highly decorated with frets. The stool clock had a simpler link between verge and crown wheel escapement and pendulum than the English lantern. This link was also devised by the inventive Huygens. A wheel balance is fixed to the verge, which is vertical. The Dutch merely replaced the wheel by a rod which connected with the pendulum.

The English longcase clock was very closely copied by the Dutch, however, except that in Holland a bow front to the plinth at the bottom of the case was often favoured.

The Dutch also set the long pendulum movement in a wall clock, the *staartklok*, that was unique. The dial was painted, like those of most provincial English longcase clocks, but there the resemblance ended. The trunk was thin, just thick enough to take the seconds pendulum. The weights were exposed and hung in front of the trunk. There was, of course, no plinth or foot.

Animated displays are a feature of many Dutch clocks, a common one being little people striking bells. The Dutch, unlike most other nations, had no inhibitions about repainting dials or automata.

In the late seventeenth century, in the reign of Louis XIV, the French introduced a bracket clock with a short pendulum, which became known as a *religieuse* because of its severe style. Compared with the really austere English bracket clock of the same date, it was quite florid, with marquetry-decorated case and applied metal frets below the dial and quite often to the case as well. The amount of decoration varied with different makers, and the name seems to refer more to the plain rectangular shape of the wooden-cased clock than to its decoration.

It was the French who really transformed the simple longcase into an elaborate, sometimes hardly recognizable, confection of marquetry and rococo ornamentation, often making the clock appear like a bracket clock standing on a pillar, although the pillar contained the long pendulum. They also diverged from the English in preferring a pin-wheel to an anchor escapement. The pin-wheel escapement, like the anchor, but unlike the verge, allowed a small arc of pendulum, which reduced errors in time-keeping and meant that longcase clocks could be made with relatively narrow trunks.

Left : Dutch Friesland alarm timepiece with a single hand and weight drive, and a verge and pendulum. The bracket hood and baseboard are replacements.

Right : An unknown German maker produced this rack clock. The clock is driven by its own weight as it falls slowly down the toothed rack. It is wound by simply raising it.

Left : Ceramic cases were popular at times, particularly in France. This one houses a clock made by Jean Baptiste Dutertre in about 1760, in Paris.

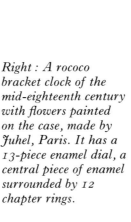

Right : A rococo bracket clock of the mid-eighteenth century with flowers painted on the case, made by Juhel, Paris. It has a 13-piece enamel dial, a central piece of enamel surrounded by 12 chapter rings.

French provincial makers enthusiastically took up the longcase clock and, instead of following leading makers in the capital city, as happened in England, went their separate ways and developed several recognizable styles, all plain, or relatively so. Some cases were straight-sided from top to bottom with simple or no decoration. In some regions, the biggest emphasis was laid on the pendulum, which was covered by a large violin- or pear-shaped piece of thin-gauge metal, decorated with repoussé work and bright colours and showing through a large glass panel in the door of the trunk, which was also violin- or pear-shaped.

Iron movements for provincial clocks, with pendulum and verge escapement, or an unusual straight rack-striking arrangement, were made in huge quantities in the Morbier-Morez-Foncine region from the eighteenth century. They are still made there today with anchor escapements.

The dials of most French clocks were round and enamelled, those of provincial longcase clocks being set in repoussé-decorated shields, often of oval shape. French leading makers had pioneered the enamelled dial both for clocks and watches, while makers of other countries kept to brass which was gilded or silvered, or occasionally used silver. Because it was difficult to enamel a large area, the early enamelled dials were made up of 12 separate pieces, one for each chapter (the hour numbers), or 13 pieces, the extra one for the centre.

The French were most prolific in case design. One that was frequently copied and lasted for centuries was the so-called balloon shape. It first appeared in the Louis XV period (1715–74). It was later, however, when the Montgolfier brothers caught the public imagination by their experiments with rigid hot air ballons, that this clock shape began its run of popularity.

The idea caught on in England, where the case was much less decorative and the shape more exaggerated; in fact more like the Montgolfier balloon in outline. The Swiss became so enthusiastic about it that they adopted it and make it to this day under the name of Neuchâteloise.

Another French design, called the *cartel* clock, although why is not clear, was an ornate wall clock in a carved case based on scroll and foliage, usually acanthus leaf designs. The case appears to be made of bronze, and indeed earlier ones were, but

Above right : An urn clock, a popular French style, made in Paris towards the end of the eighteenth century by Dufaud. The height is 73cm (2 ft 4 ins). The band with the hour chapters revolves against a fixed hand.

Far right : The elaborate French clock on an elaborate bracket were sometimes combined to become a single cartel or wall clock. In this piece, made about 1770, it is impossible on first sight to say whether it is in one part or two.

often it is a clever imitation of bronze, invented by the Martin brothers in 1730, and known as *vernis Martin*. The carving was of oak, treated with gesso to hide all cracks and grain, then polished, coloured, and varnished to make it appear to be of polished metal, although sometimes non-metallic colours were also used. After some years, the surface usually becomes covered by a network of tiny cracks. Swedish makers also favoured the *cartel* style.

Many beautiful marquetry cases were made by the French, and not only using different woods, as the English and Dutch did, but often with brass and tortoiseshell, a finish invented by Charles André Boulle (1642–1732). The preparation of a Boulle case was extremely demanding of the artist-craftsman, as the very complex pattern was often applied to curved and even convex surfaces.

The greatest age of ornamental French clockmaking was in the second half of the eighteenth century, during the reign of Louis XVI. French clockmakers had overtaken the English by mid-century, although the English had taken the lead in watchmaking (which they were eventually to lose to the Swiss). It was to be the intervention

Left : A neoclassical design in a composition of musical instruments carved in lime wood, gilded in red and green and partly silvered. The height is 75cm (2 ft 5 ins). It was made in Prague about 1790 possibly by Ignaz Michal Platzer.

of Louis XVI, from 1778 to 1783, in the American War of Independence that led indirectly to American domination of clock-making before the mid-nineteenth century.

The Louis XVI period was the finest for bronze or bronze and marble clocks with figures, foliage, and often enamelled panels. The human figures were usually classical Greek or legendary. Classical columns, vases and harps were also commonly part of the design.

Original clocks of the period are now not easy to find. The large numbers seen in antique shops are virtually all copies of Louis XVI clocks and those of the Empire period that followed, made during what has been called the Second Empire Period from about 1852 to 1870. The cases were reproduced by making electrotypes backed by spelter, which were then fire-gilded, using a mixture of ground gold (or *moulu*) and mercury. This butter-like mixture was rubbed on the case and the mercury then burned off, a noxious process dangerous to health and not now used. This means that later reproductions and restorations are electro-plated, which gives a different finish.

Other nations, including the English and the Germans, copied French styles, in some

Left : An Empire clock in the form of a charioteer, in ormolu, with the dial in one wheel. It was made by Gentilhomme, Paris, c. 1810, and the height is 45cm (1 ft 5 ins). The chariot is decorated by the figure of Athene and a lion's head.

cases so closely that it is difficult to decide the origin from the case. English-made 'French' clocks normally have much higher winding holes in the dial, which is one clue.

The longcase clock appeared in Belgium, following the English fashion of a brass dial with silvered chapter ring, but often with applied decoration or carving on the case. Germany made them, too, but more in the French provincial style, as round enamelled dials were favoured. Cases were plain and unfussy, sometimes having parquetry patterns, but were not so classical in style as the English.

The Scandinavians, too, took to the longcase clock, Danish makers following most closely the original English styles in both dials and cases; some of their clocks could have been made in London. That did not prevent some Danes from venturing into more Dutch-like designs, however. The Swedes and Finns were more influenced by the French.

BLACK FOREST CLOCKS

By the nineteenth century, the second of the two great pioneering clockmaking towns in Germany, Augsburg, had waned, but another centre had grown up without any historical

Left : A number of countries copied the English longcase clock. This strange looking version was made in south Germany by an unknown maker about 1750. It has a long pendulum and strikes the hours.

Right : A Belgian version of the longcase clock, made about 1750 by Gilles de Beefs, Liege, in the English style. It has an anchor escapement and long pendulum and the lever in the dial arch is to silence the striking.

Left: An early Black Forest hanging clock of about 1760 with a chime on bells. The 'cow's tail' pendulum hangs in front of the dial. The movement and its parts are of wood and the bells of glass.

Right: An Austrian clock (c. 1790) by Paulus Hartmann of Vienna, that runs for a month at a winding. The movement and dial form the bob of the seconds pendulum, which has a Harrison gridiron temperature compensated pendulum rod.

Left: A French baroque clock of the eighteenth century made in Paris by Nicolas Delauney, in the reign of Louis XIV when enamelling a whole dial was very difficult, so the dial was made of 12 enamelled plaques, like this, then later with a thirteenth in the centre. The case is of green varnished wood with five gilded bronze ornaments.

connection with the past. It was in the Schwarzwald, the Black Forest, in Baden and Württemberg, in the south-west.

Areas of the forest had been cleared of trees for farming. The farmers had little to do in the winter because they and their livestock were confined to wooden houses while snow covered the land. So some of them, using the plentiful wood, became skilled at carving toys and souvenirs for visitors.

One day, around 1670, one of the carvers attempted to copy a simple iron clock he had obtained – and it worked. From that unlikely start, the world's biggest clockmaking industry grew. At first all Black Forest production was craft-made in wood. The frame and plates were wooden, as were the wheels and arbors, and the cases and dials.

The first design was based on a watchman's clock of the fifteenth century, which was weight-driven, had a verge and foliot, and hung on the wall. The Black Forest version had two strips of wood, united at top and bottom by two much shorter strips to form the frame. There was a single wooden hand. A stone provided the driving weight. The problem of making a crown wheel in wood was neatly side-stepped by using a wooden disc with a row of pins projecting called it a *Kuhschwanz*, meaning a cow's tail. For this a crown wheel had to be carved in wood. The pendulum bob was also of wood, sometimes with a brass facing. The long pendulum with anchor escapement was introduced not many years later, but the cow's tail lingered on for a number of years.

The traditional Black Forest clock, after the introduction of the long pendulum, became a wall clock known as a *Schild*, which was basically a boxed-in movement with an exposed dial and weights and pendulum hanging below it. The dial was colourfully decorated and much preparation went into it. The round, rectangular, or arch-topped wooden dial plate had a ring of wood glued to it. The ring was turned on a form of lathe to make it a convex chapter ring. The whole was then smoothed with tripoli and powdered

Above : The later form of cuckoo clock case, made after 1850. This particular one is a cuckoo-quail clock, in which the second bird automata sounds the quarters.

Right : A Black Forest clock (1710). It is in a wooden case and has a wooden movement, including the wheels and arbors, and a foliot and two glass bells, the lower one for the hours, which are struck by miniature jacks. The two dials show hours and quarters.

from one side near the edge.

Glass objects were also produced in the Schwarzwald in the seventeenth century, so it is not surprising to find early wooden clocks with glass bells for striking. When stones were abandoned for weights, glass tubes filled with sand or shot took their place, until displaced by cast iron. Even musical clocks with a series of glass bells were made. By now the elementary frame had been supplanted by two wooden plates separated by four pillars, just like a brass-plated movement.

About 1740, Christian Wehrle introduced the pendulum to Black Forest clockmakers. It worked with a crown wheel and verge and swung in front of the dial, in the same way as the pendulums of earlier clocks which had been converted. The farmers pumice stone and a white background was painted on before the decorative designs and final varnishing.

Black Forest makers had built up a big export business before the end of the eighteenth century and were making many other designs to suit their customers. For Britain, they made clocks with circular dials and wooden rims, generally similar to the clock called an English dial, which was a round spring-driven wall clock. The German one was weight-driven, and early versions had no glass in front of the dial. The so-called 'postman's alarm', many are still seen today, was one such clock.

Around 1730, a Schönwald maker, Franz Anton Ketterer (1676–1750), had the extraordinary idea of using the call of the cuckoo for the striking sound in a clock. No one knows why. There was no striking or chiming work in any of the earlier Black Forest clocks, so perhaps, with no tradition to influence him, Ketterer took an obviously alerting call from nature.

Complete movements were made of wood throughout the eighteenth century, but pivots of wire, inserted in the ends of the wooden arbors, were found to produce less friction; wire was also used to produce lantern pinions. It was not until the next century that brass replaced wood for wheels. The first wheel to be replaced was the escape wheel and, in 1790, one clockmaker speeded up the process of manufacture by casting them. Plates continued to be made of wood until the 1840s.

The industry soon separated itself from farming and towards the end of the eighteenth century there were almost 1000 master clockmakers, supported by workers in various ancillary trades, who produced nearly 200,000 clocks a year. The major problem was not production, but selling the clocks. In the early days, the clockmaker employed glass salesmen. A salesman would set out dressed in traditional knee breeches, long white stockings and broad-brimmed hat, with a pack of clocks on his back to sell in other villages and towns. The salesman would sound one of the clock bells from his *Tragstuhl* (pack) with his fingers to attract attention. Naturally, carvers also made models of the salesman with his *Tragstuhl*, usually with a working clock mounted in his chest.

Soon the travelling salesmen were going to other countries and they continued to do so well into the nineteenth century, although clockmakers had meantime organized themselves into factories and had also adopted other methods of selling, including a not very satisfactory sale-or-return or 'packer' system, so-called because the packer, often a shopkeeper or innkeeper, took orders, ordered packs of clocks from clockmakers, then collected the money from sales. The clockmakers had no means of knowing what had or had not been sold, because unsold clocks could accumulate on shelves and they could go on making for a market that no longer existed.

Sometimes today, an antique dealer is heard to refer to a 'Dutch clock'. Invariably the clock is from the Black Forest. Charles Dickens and other writers described certain clocks as 'Dutch'. The mistake originated when the travelling salesmen from the Forest described themselves as *Deutsch*.

Below: A German weight-driven shield clock, a popular Black Forest style of hanging clock, with colourful wooden dial and weights hanging below it. This one is also a cuckoo clock.

V
THE TIME AT SEA

The position of the sun was not only man's first timekeeper, but also his first means of estimating his position when trying to navigate a boat out of sight of land. Despite much of what has been written, early sailors did not hug the coast and risk their lives and boats on the rocks and sandbanks of unknown territory.

Homer, in describing the adventures of Odysseus after the fall of Troy, explained that Poseidon, God of the Sea, 'went for a visit a long way off, to the Ethiopians, some near the sunset'. The shadow of the sun at noon points north and south. At right angles to this meridian line are the directions east and west. The Greeks added new directions, those of sunrise and sunset in summer and winter. Each was about a third of the way round the original quarter division, so the quarters were divided into three, and the early compass had 12 points like the dial of a clock.

Before then, Greek mariners used the prevailing winds, which became the wind rose, which was in use in the Mediterranean for a thousand years or so before the compass rose. The Tower of the Winds, the elaborate hydraulic clock in Athens, identifies eight of these winds.

There is no doubt that in ancient times sailors were able to navigate at night. Before 700 BC, Homer described Odysseus as navigating by the stars. This was done by watching the Great Bear (Big Dipper), as it circled continuously around the celestial pole, like the hand of a heavenly 24-hour clock, but in the northern hemisphere, going backwards.

The navigators of ships in the fifteenth century were 'learned men; astrologers and watchers of omens who considered the signs of the stars and judged the winds and gave directions to the pirate (pilot) himself'. The star-watchers knew by now that the stars they used as clock hands gradually went more slowly, so that every two weeks they were an hour out. Some time before 1520, however, a new 'clock' was invented,

Far left: John Harrison, painted in oils by Thomas King, (1750). He is holding a small watch, probably the one which was made in 1753 to try out the ideas Harrison used in his famous number four marine timepiece. Behind on the right is an inaccurate portrayal of a gridiron pendulum.

Left: Sandglasses were commonly used as interval timers on ships for measuring speed in navigation. A log was thrown over the side with a knotted line attached to it. The speed at which the knots ran out was measured by the glass. This one, in Madrid, was made at the end of the sixteenth century.

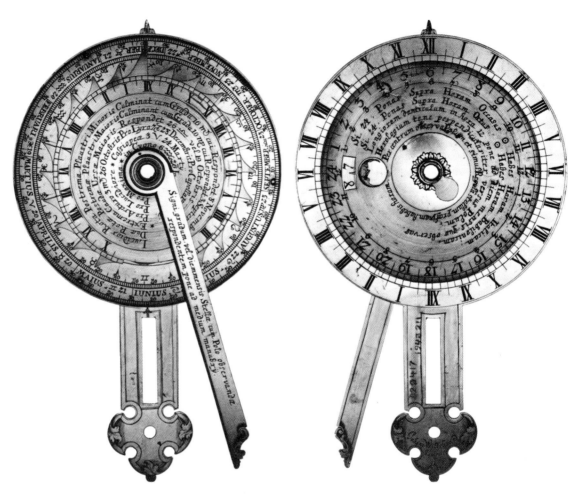

to tell the time at night by either the Great or the Little Bear.

It was called a nocturnal and was a wooden or metal disc with a hole in the middle, held to the eye by a handle, like a lorgnette. A short arm was moved to the date on an outer scale, and the Pole Star was viewed through the hole. A longer arm was lined up with the guards or pointer stars of the Bear, and this indicated the time on an inner scale. The time was read off on the inner scale by taking it inside to the light.

From the end of the sixteenth century, the most common way of finding the time at sea by night was by the nocturnal, and during the day by the universal ring dial, a more sophisticated version of the simple sundial.

The ring dial comprises two rings at right angles, one horizontal and the other vertical. Across the vertical one is a bar, the polar axis, which carries a slide with a hole in it. The slide is moved to the current date, marked on the axis. The ring dial is then held by a suspension ring at the top until a spot of sunlight through the hole falls on a scale on the horizontal ring, where it indicates the time. If this is local time, the vertical ring also indicates north and south.

THE LONGITUDE PROBLEM

Perhaps the greatest step in the art of navigating in Western civilizations was made by the astronomer Ptolemy, who, in his work *Geographia*, written in about AD 150, introduced the ideas of latitude and longitude. But it took from then to the Renaissance for these imaginary lines to be used in practical navigation. By then, armed with a chart on which lines of latitude were marked, a mariner could take sightings of the sun and stars to locate his position with reasonable accuracy north or south.

Finding a ship's longitude – its position east or west – however, was a different matter. The problem baffled the greatest brains in astronomy, mathematics, physics, horology and navigation for centuries. Losses at sea were reaching the magnitude of national disasters. One of the worst for England was when Sir Cloudesley Shovell was returning with his fleet from Gibraltar in 1707. Altering course to run before a westerly gale, the fleet found itself suddenly among the rocks of the Scilly Isles at night. Four ships with about 2000 men and the Admiral himself were lost.

Sea-power was so vital to rival trading nations that several governments offered

Above : A nocturnal was used for finding the time at night. Here are both sides of an elaborate version made about 1700 by Johann Willebrand, Augsburg. The ornate arm is the handle and the other the sighting arm. The hour ring is set to the hour by the toothed ring. On the back is a silver plate to convert Italian to Babylonian hours.

substantial rewards to anyone who could discover a means of plotting the longitude. In 1598 Philip III of Spain offered a perpetual pension of 6000 ducats, a life pension of 2000 ducats and a gratuity of 1000 ducats to anyone who could find a solution to the problem. Christiaan Huygens, the astronomer and inventor of the pendulum clock, referred in his writings to an award offered by the States-General (now Holland), estimated to have been 1000 to 30,000 guilders. France offered 10,000 livres as a Prix Rouille in 1715, and Venice also offered a prize. The British Government made the biggest offer of all.

As a result of a petition to Parliament about the seriousness of the longitude problem, an Act was passed in 1714, in Queen Anne's reign, 'for providing a Publick Reward for such Person or Persons as shall Discover the Longitude at Sea'. £10,000 was offered for any method of determining longitude to within one degree, £15,000 to within 40 minutes of arc and £20,000 to within 30 minutes of arc, that is, about 50 kilometres (30 miles) at a latitude just north of the equator.

A body of commissioners known as the Board of Longitude was set up under the Act to examine proposals, and during its existence from 1714 to 1828 paid out the huge sum of £100,000, the equivalent of millions today. The Board had to deal with a number of cranks, as may be imagined. The minutes of 25 January 1772 record: 'A person who calls himself John Baptist desiring to speak with the Board. . . .'

Nearly a century before the British Government's offer, the astronomer and

Left: A universal ring sundial made in the eighteenth century in Germany. It has a pin gnomon, the shadow of which indicates the time. The hour circle with gnomon can be angled to the local latitude, but is not suitable for use at sea, as it has to be levelled and orientated, for which a plummet and compass are provided.

scientist Galileo had made an unsuccessful bid for the Spanish prize. Years later, he offered the idea to the States-General, again without success. After making his astronomical telescope in 1609, Galileo had discovered that the planet Jupiter had satellites. He realized at once that, as the satellites were eclipsed almost every night, they should be better than the moon for calculating longitude. Armed with a table of times when the eclipses occurred in relation to a particular meridian (line of longitude), a navigator at sea could note the difference between this and local time on the ship from a sun or star fix and calculate his longitude, allowing one degree of longitude to four minutes of time.

Apart from the astronomical observations needed to compile accurate tables, the

Left: A folded universal ring dial suitable for use at sea. The equatorial ring (inner one with Roman numerals) is turned at right angles to the outer meridian ring. The bridge (crosspiece) is turned flat and to the correct latitude, and the slide with pinhole moved to the correct date. When the instrument is suspended from a cord and a spot of sunlight falls on the equatorial ring, the spot shows local time.

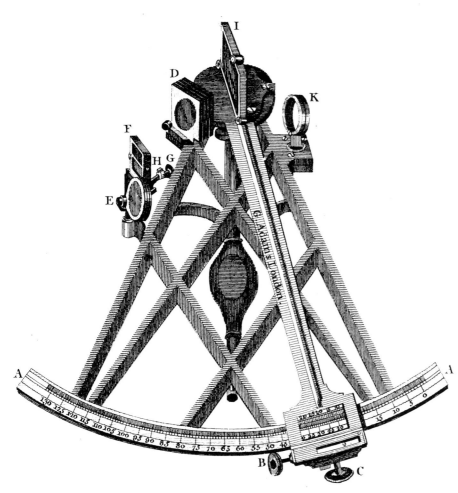

difficulties involved in aiming and focusing a lengthy primitive telescope on board a tossing sailing ship at a tiny object in the sky were, to say the least, formidable. Even that was only part of the problem, because a good timepiece was still needed to time the observation. After the foundation of the Paris Observatory in 1672, much study was made of Jupiter's satellites, and from about 1690 Galileo's method was used extensively, but only on land, where it was used to find the longitude when drawing maps.

Another suggestion, first made by Regiomontanus in 1474, and revived by the German star-gazer John Werner in 1514, was to use the moon as a clock by measuring its motion. It seemed an eminently reasonable idea, but turned out to be impracticable, so a more complicated method known as 'lunar distances' was devised.

The method of lunar distances was most difficult, and remained so even after centuries of effort to improve it. No modern navigator would dream of accepting the calculations involved. They entail measuring the distance between the moon and the fixed stars, including the sun. This depends on knowing the moon's position in advance. Even today

Above : The quadrant was improved gradually, mainly by John Hadley, who, in 1723, provided two mirrors, one to bring the image of the heavenly body being observed to the horizon, which had been aligned on the other. The height was read off the scale.

Right : The Royal Observatory in Greenwich Park, showing a quadrant (left) and a telescope. The two clocks set in the back wall on the left were made by Thomas Tompion. Both ran for a year at a winding and had 4.27m (14ft) pendulums, beating two seconds, above the the movements.

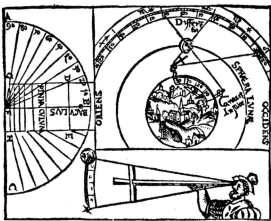

Left : The cross-staff – first described in 1342 – in use to make observations of the moon to obtain local time. The illustration is from Cosmographia *by Gemma Frisius (1542).*

it is impossible to predict this five years ahead, owing to the irregular rotation of the Earth, which makes the moon behave like a bad clock.

Nevertheless, belief in the method of lunar distances was so strong that it was a principal motivation behind the setting up of the Royal Observatory at Greenwich by Charles II, three years after the French had completed their observatory in Paris. A young Derbyshire clergyman named John Flamsteed (1646–1719) had been working on the moon's motions and was asked by a committee set up by the King, which included the great experimenter Dr Robert Hooke of the Royal Society, to report on a scheme, for using the heavenly bodies as an aid to navigation, put up by a Frenchman. This he did so successfully that the King set up a Royal Observatory at Greenwich, and made Flamsteed the first Astronomer Royal. However, the royal purse was kept so tight that Flamsteed had to depend largely on gifts, both to buy instruments and for his own keep. The Observatory itself had to be built with old bricks. Christopher Wren designed it, and the King raised the money by selling 'old and decayed gunpowder'.

Flamsteed was charged with applying himself 'with the most exact care and dilligence to the rectifying the table of the motions of the heavens, and the places of the fixed stars, so as to find out the so much desired longitude of places for the perfecting the art of navigation'. He grew old at the task and his star catalogue was not published until after his death.

The fifth Astronomer Royal, Nevil Maskelyne (1732–1811), who was also a clergyman, eventually produced the requisite figures for the method of lunar distances, and he tested them himself at sea, publishing his findings in the *British Mariner's Guide* in 1763. One navigator reported that he had used the methods prescribed by Maskelyne, which 'were found very useful and not difficult, each observation not taking more than four hours to find the result'. Four years later, Maskelyne instituted the celebrated *Nautical Almanac*, which published lunar distances regularly.

Even a laborious four-hour calculation could not have been made without the invention of two instruments. One was Edmond Halley's reflecting quadrant to take the sun's angle for local time, and the lunar distance for time at the standard meridian. The other was a timepiece accurate enough to make a correction between the times of the two observations.

PURSUIT OF ACCURACY

To give a position to about 100 kilometres (60 miles) after a six weeks' voyage, a watch would have to keep time to within six seconds a day. This degree of accuracy was in the realms of fantasy at that time, and many eminent men who sought solutions had to admit defeat. They included Sir Isaac Newton, Sir Edmond Halley, Christiaan Huygens, G. W. von Leibnitz and Robert Hooke.

Christiaan Huygens (1629–95) was the

Above : John Flamsteed, the first Astronomer Royal, was appointed at the age of 26. He was an extremely conscientious man who suffered from poor health and spent a lifetime compiling an astronomical almanac.

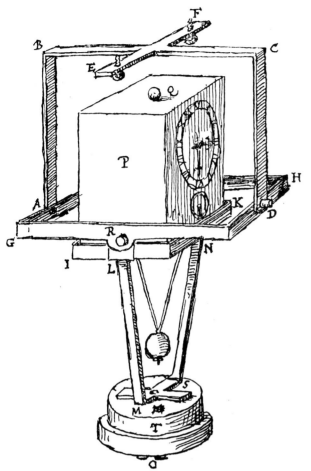

Left: A sketch made by Huygens in about 1660 of his sea clock. Encouraged by Alexander Bruce, Earl of Kincardine, an exile in Holland, he became interested in the problem of the longitude, but had no success with two small pendulum clocks. Then he designed this one held in gimbals and steadied by a weight below it. The pendulum was triangular with cycloidal cheeks and a remontoire, but it was not successful either.

oscillated to control the rate. After the addition of the spring, watches would go accurately enough on board ship to transfer a time over a period of a few hours. However, they were still not accurate enough for time-keeping over several days to find the longitude directly.

A Yorkshire clockmaker named Jeremy Thacker followed Huygens in a determined assault on the problem. He wrote a pamphlet lampooning contemporary efforts to solve the problem and putting forward a serious proposal of his own. It was entitled *The Longitudes Examined, beginning with a short epistle to the Longitudinarians and ending with the description of a smart, pretty Machine of my Own which I am (almost) sure will do for the Longitude, and procure me The Twenty Thousand Pounds.* It was published for sixpence in 1714. He described a spring for keeping the clock going while it was being wound, and perhaps it is he, rather than John Harrison, who should be given the credit for this idea, called maintaining power. He also suggested a clock in a vacuum, but he had been preceded in this idea by 'the accurate clock of Antonia Tempera, most useful to navigators', which was enclosed in a Torricelli vacuum and was described in a book of 1668.

Thacker ended his pamphlet with the sentence: 'In a word, I am satisfied that my Reader begins to think that the Phonometers, Pyrometers, Selenometers, Heliometers, and all the other meters are not worthy to be compared with my Chronometer.' He appears to have been the first to use the word chronometer for a high-precision timekeeper.

An English clockmaker named Henry Sully, who spent all his working life in France, made a determined attempt to solve the problem with timekeeping machines that were highly ingenious and beautifully made, but had some fundamental faults. Their success depended upon a horizontal weighted bar instead of a hairspring. Unfortunately, when the ship rolled or pitched the weighted bar displayed all the faults of a pendulum. Sully made a useful contribution by introducing roller bearings for marine clocks, an innovation which a number of other makers followed.

first to build a clock intended solely for finding the longitude. His invention of the pendulum clock had revolutionized the accuracy of timekeeping on land, so he proposed to use a pendulum at sea. In 1660 he produced a sea clock, the first ever designed for the purpose. It was suspended from gimbals – a form of universal joint – so that it remained horizontal no matter what tack the ship was on and whether or not it was pitching.

The pendulum was a triangular arrangement with the bob at the bottom point, and the clock was spring-driven. Although it worked in calm seas, as soon as the ship started to roll the pendulum would stop and restart, or jerk, and it would behave erratically when gravity changed as the ship rose and fell. Huygens's reputation carried the idea a long way, and his timekeepers were tested by the French navy. Unfortunately, they were useless.

Huygens then turned his attention to the watch, which was a very poor timekeeper, but at least kept going at sea. In 1675 he made a remarkable improvement in its timekeeping by adding a spiral spring to the oscillating balance wheel. Up to then, the watch only had a wheel or a bar which was

JOHN HARRISON

It took a century of experiments, from 1660 to 1761, before a timekeeper was made that could keep time accurately enough at sea. Despite the fact that the greatest intellects of the age and of all civilized nations were concentrated on the problem, it was solved by John Harrison (1693–1776), a carpenter and joiner from the small Lincolnshire village of Barrow on the southern bank of the river Humber, who had no formal education and not even an apprenticeship in clockmaking.

Harrison was interested in making clocks from an early age, and by 1726 he and his brother James had made a few precision longcase (grandfather) clocks with wooden movements and wheels that were so brilliant in conception and execution that they were credited with keeping time to within a second a month. They incorporated an original escapement and bearings that eliminated the need for oil and reduced friction to a minimum. Even more important were their devices to obviate the effects of changing temperature on timekeeping. The clocks still exist, one belonging to the Worshipful Company of Clockmakers and the other is in private ownership.

It is incredible that these two carpenters were able to design and make horological machines so scientifically sophisticated in a little village cut off from centres of intellectual activity. When he was old, John Harrison declared that he did not hear of the

Right : The dial and hood of a longcase clock made by James Harrison, Barrow, c. 1726, with black and gold decoration. The apertures show seconds and the day of the month. There is an equation of time table mounted on the front of the case.

Below : A marine timepiece made by Henry Sully in France c. 1724, which fitted into a burr walnut case like a bracket clock. A balance suspended on anti-friction wheels was attached to a horizontal pendulum with cycloidal cheeks. It was not successful.

Act of 1714 until 1726, that is, until the precision clocks had been made, but this must be considered unlikely: the big port of Hull was not far off, and the enormous value of the prize must have been common knowledge there. The only solid fact known about John Harrison's education is that a visiting clergyman was so impressed with his intelligence that he lent him a manuscript copy of a series of lectures on natural philosophy by Nicholas Saunderson, the blind professor of mathematics at Cambridge University. Harrison made a meticulous copy of the lectures and diagrams in about 1713, and consulted it in later life, as marginal notes indicate.

How did John Harrison regulate his precision timekeepers to such accuracy of rate, and how did he check the rate? He himself explained, in a manuscript written in 1730. For two nights he watched a fixed star at night align itself with the west side of a window frame in his house and the east side of a neighbour's chimney-pot. A helper counted down seconds on one of the clocks until the moment the star disappeared behind the chimney-pot. The star gave sidereal time, or star time, which is 3 minutes 56 seconds shorter than clock time over 24

Harrison himself recorded that 'Mr Graham began as I thought very roughly with me, and the which had like to have occasioned me to become rough too. . . .'. Once they had sized each other up, they got on well. Graham advised Harrison to return home to build his clock. Moreover, he arranged to finance the work, by an interest-free loan from himself, and, over a period, £100 from the East India Company and £80 from a Mr Charles Stanhope.

What Harrison did not realize was that the building of a successful sea-going time-keeper would, in fact, take him another 30 years. The brothers' first sea clocks were large machines. Number one weighed 33 kilogram (72 lb.) and was 90 centimetres (3 feet) in all dimensions. It incorporated two large interconnected balances – centrally

Above: Edmund Halley was the Astronomer Royal who first saw John Harrison in London, and sent him to see George Graham. Graham advised and financed Harrison, despite some clash of character.

Right: Harrison's No. 1 'sea clock'. Hours are shown on the right hand dial and minutes on the left, seconds at the top, and day of the month at the bottom. This and H.2, H.3 and H.4 are in the National Maritime Museum at Greenwich. Two large dumbell balances are interconnected by coil springs.

hours, which had to be allowed for.

Having completed this manuscript setting out his ideas and inventions and providing evidence of his ingenuity, John Harrison made a journey to London, his first, to solicit help from the Board of Longitude for his project of building a sea clock. It is likely that the unknown Lincolnshire carpenter had some difficulty in penetrating the domain of the Commissioners, but one certain fact known about Harrison is that he always refused to accept defeat. He managed to obtain an interview with the Astronomer Royal, Edmond Halley, who was not only one of the Commissioners, but was then engaged in the unrewarding task of trying to compile a table of lunar distances.

Halley found Harrison a difficult person to deal with and to assess. He told him to go and see 'Honest George' Graham (1673–1751), the famous London clockmaker and instrument-maker who was also a Fellow of the Royal Society and whose opinions were respected by the Commissioners of Longitude. Harrison protested that his ideas would be pirated by another clockmaker, but Halley assured him of Graham's honesty. The subsequent interview with Graham was none too smooth.

pivoted arms with a heavy ball on each end. The brass balls of one balance were connected by helical springs to the balls of the other. Various devices from the precision clocks were also incorporated.

Making and testing the clock took nearly five years. The Harrisons made another approach to the Board via Halley and Graham, but it was the Royal Society that examined it and issued a certificate commending it for an official trial. The machine, now known as H.1, was, on the authority of the First Lord of the Admiralty, Sir Charles Wager, given a test at sea attended by John Harrison.

The captain of the ship on which Harrison departed for Lisbon wrote to the First Lord to say that he found Harrison sober, very industrious and modest, 'so that my good

Wishes can't but attend him; but the Difficulty of measuring Time truly, where so many unequal Shocks, and Motions, stand in Opposition to it, gives me concern for the honest Man, and makes me fear he has attempted Impossibilities. . .'.

H.1 performed well, however, and the Board met, for the first time in its 20 years' existence, to consider it. Strangely, Harrison, instead of asking for a longer trial in accordance with the Act, tamely said that he was not satisfied with the performance and suggested he provide a smaller version, which would take him another two years.

In fact he made another, H.2, then another, H.3, with financial assistance from the Board. In 1755 he wrote to the Board to say H.3. was ready but that he was making two watches, a small one for the pocket, and a large one which he thought might be used to find the longitude at a cheaper cost than his large machines.

This large watch, H.4, is probably the most remarkable timekeeper ever made. In its silver outer case, it is about 13 centimetres ($5\frac{1}{4}$ inches) in diameter. It can be seen today, with the rest of Harrison's seagoing timekeepers, at the National Maritime Museum, Greenwich. John Harrison (his

Above: Harrison's second marine timekeeper H.2, completed about 1739. Hours are shown in cut-out segments at the bottom of the dial and minutes by the hand at the top. At the bottom right are anti-friction wheels.

Left: The third marine timekeeper H.3. It has two large interconnecting circular balances. It incorporates the first bimetallic strip or curb, which compensated for temperature errors.

brother receded into the background, although he worked on the timekeepers) told the Board in 1761 that both the clock, H.3, and the large watch, H.4, were ready for sea trials.

As John Harrison was now 67 years old, it was agreed that his son William should take his place on the sea trial of the watch. The watch was taken to Jamaica for the trial. On the ninth day, William Harrison calculated that they would sight Madeira the next day. The captain bet him five to one he was wrong. Harrison was right, and the ship's company was much relieved 'as they were in Want of Beer'. At Jamaica, the timekeeper was five seconds slow after applying its rate correction, which corresponded to an error of only one and a quarter minutes.

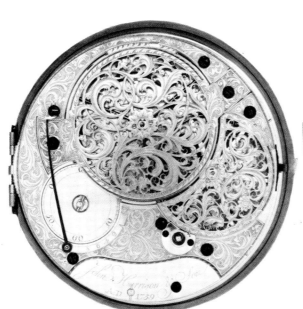

Right : The dial and top plate of the famous H.4, signed 'John Harrison and Son AD 1759', which won the British government prize of £30,000. The silver case is 13.3cm (5¼ ins) in diameter. The watch, and its prototype made by John Jefferys under Harrison's instruction, is fully jewelled, including diamond pallets to the special verge, maintaining power, a remontoire, and a centre seconds hand.

Right : A page from mechanical notes and drawings by John Harrison and his son William now in the Clockmakers Company museum at the Guildhall, London. It shows the layout of the wheel work of H.4. The notes were made between 1726 and 1772.

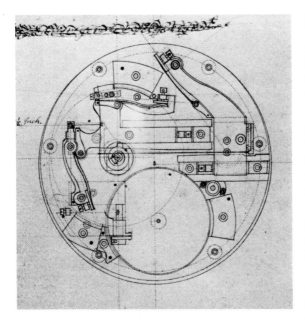

The return journey was so rough that William Harrison wrote that the watch 'received such shocks from the breaking of the waves under the Counter, that it was just as if I had taken the box in my hand and thrown it from one side of the Cabbin to the other'. Storms on the way made water 'spurne in at every joynt and we sometimes had two feet of water on our decks and 6 inches deep in the Capn. Cabbin'. Harrison had to wrap the box containing the time-keeper in a blanket and keep changing the blanket for a dry one.

Nevertheless, the prize of £20,000 was at last within their grasp. Or so it seemed. The Board thought otherwise. The result might have been a fluke. They offered John Harrison £2500, to be deducted from any future award, and asked that the watch be

given another trial. Faults lay on both sides. The Commissioners wanted Harrison to disclose details of H.4 in case it was lost at sea. Harrison agreed, but did not provide the information. However, he did consent to another trial.

The first result had been so fantastically good for the time that it is not surprising it was thought to be a fluke. But the Harrisons should have explained what was meant by rate correction. The Commissioners were undoubtedly muddled about this.

The rate of a timekeeper is its consistency in going. A watch that is known to gain a second a day will be 80 seconds fast after 80 days; thus it is only necessary to subtract 80 to get the accurate time of day. On the other hand, a watch may be known to vary in rate from two seconds fast to three seconds slow. As it is impossible to check all the time whether it is gaining or losing, after 80 days it could be a maximum of 160 seconds fast or 240 seconds slow. It could also be exactly right. There is no means of knowing.

On the second trial, which was to Barbados, four mathematicians appointed by the Board checked the results of the land observations and the time given by H.4 to which the known rate corrections were applied. The watch was three times more accurate than necessary to win the highest award offered by the Act. When the Board met to consider the results, they did not invite the Harrisons. They seized upon the wording of the Act that the method had to be practicable and useful at sea. How could they judge this when Harrison had not disclosed full details of his timekeeper?

Left : Harrison's last marine timepiece, H.5, which he made under protest. It is like H.4 but plainer. The gold star in the middle is for setting the hands. George II tested it in 1772.

Below : The Reverend Nevil Maskelyne, Astronomer Royal, in an oil portrait by J. Downman, butt of Harrison's comment, 'the Devil may take the Priests'.

The Board at this time had become enthusiastic about lunar distances, because a test, also on a journey to Barbados, had produced a result equal to accuracy within half a degree or 30 minutes of arc, good enough to win the full prize. The lunar method was made feasible by new tables and was strongly backed by Nevil Maskelyne, who was now the Astronomer Royal and a member of the Board.

After the first trial, John Harrison had begun publishing pamphlets about his grievances, some strongly worded. The Board now wanted him to produce some timekeepers to prove that H.4 was not a freak. It was 1765, and Harrison was 72 years old. In anger and desperation, he petitioned Parliament, but to no avail. Within three months Parliament passed a new Act awarding Harrison £10,000 as soon as the principles of his watch were explained, and another £10,000 only after other timekeepers had been made and tested. Harrison was certain that the only result would be the pirating of his ideas. At that time there was no effective way of copywriting or patenting a design, certainly not internationally.

The French made some attempts to buy Harrison's secrets from him, and apparently Harrison said he would sell them for £4000. Three specialists, including the famous French maker Ferdinand Berthoud, were sent over, but returned empty-handed. Subsequently Berthoud came again, but since the French offered only £500, Harrison turned down the offer. In the meantime, Berthoud managed to make contact with a member of the Board who promised to disclose the secrets. Harrison revealed this in a pamphlet and named the member as another famous maker, Thomas Mudge. The Board later called Mudge to account, but he claimed that the Commissioners had never said that the disclosures Harrison had made were to be kept secret.

Eventually, Harrison was forced to make drawings and to give up his timekeepers or forfeit even the first instalment of the money. The successful H.4 went to the Admiralty. The three large timekeepers he was allowed to keep until the Board called for them. Another maker, Larcum Kendall, was engaged to make a copy of H.4.

When the Board wanted the three earlier timekeepers, Nevil Maskelyne, by now Harrison's arch-enemy, called for them unexpectedly at Harrison's house in Red Lion Square, London. The aged Harrison was shocked and bitter. He wrangled about

Left : The back plate of Larcum Kendall's first duplicate, K.1, of H.4. Captain Cook used it for navigation in his second voyage to the Pacific in 1772.

Below : Kendall's third watch, K.3, was practically identical to K.2 except for the escapement and the three separate dials. It was set in a case in gimbals and Cook used it on his third voyage on Discovery.

a receipt, wanting Maskelyne to sign that all the timekeepers were in perfect order. Maskelyne refused. So Harrison would not explain the best way of transporting the timekeepers until the anger of Maskelyne and a Dr Long, who had arrived meanwhile, forced him to do so. The workmen then dropped H.1, and Harrison claimed that all the timekeepers were damaged in transit.

Maskelyne carried out on H.4 some trials that were very rigorous and in some ways unfair. His unsatisfactory results brought him into conflict with Harrison again.

Meantime, not only was Larcum Kendall carrying on with his copy of H.4, which became known as K.1, but Harrison and his son were themselves making a copy, a very plain version of H.4, now known as H.5. They completed it when John Harrison was 77. The watch was tested by 'Farmer George' – King George III himself – in his private observatory at Richmond, but it behaved erratically. The reason was found to be some lodestones in a neighbouring cupboard.

The Harrisons, feeling grossly cheated by the Board and by Parliament, continued pamphleteering and petitioning and eventually a debate was held in Parliament. The famous Edward Burke pleaded on John Harrison's behalf. Eventually an Act was passed giving John Harrison £8750, making up the total to £18,750, which provoked William Harrison into stating that his father had been stung out of £1250. Harrison had, however, during his life received various grants amounting to £4000.

The final money was paid in 1773. Harrison still continued writing pamphlets.

Left: The dial and top plate (far left) of K.2, in which Kendall omitted Harrison's remontoire and lowered the performance. Captain Bligh, of the Bounty, was using it for navigation at the time of the mutiny and it was taken by the mutineers. On the top plate can be seen the bimetallic compensation curb, which lengthens or shortens the balance spring according to the temperature.

In his last, in 1775, he made his final thrust:

> . . . if it so please Almighty God, to continue my life and health a little longer, they the Professors (or Priests) shall not hinder me of my pleasure, as from my last drawing, viz, of bringing my watch to a second in a fortnight . . . And so, as I do not now mind the money (as not having occasion to do so, and withal as being weary of that) the Devil may take the Priests . . .

The next year he died.

The final irony of the sad but inspiring history of the natural genius and doggedness of John Harrison is that, despite its feats of timekeeping, H.4 had little effect on the development of the true chronometer.

Larcum Kendall's K.1 was taken by

Below left: Larcum Kendall was a superb craftsman, but not a successful innovator. This pivoted detent pocket chronometer of 1786 was his only attempt at a precision watch of his own design.

Captain Cook on his famous journey south when he crossed the Antarctic Circle for the first time in history. Captain Cook reported that 'Mr. Kendall's watch has answered beyond all expectations', and referred to it as 'our never failing guide, the Watch'.

The Board wanted Kendall to make more watches like Harrison's, but Kendall declined because of the time it would take. He was persuaded to make two more of a simplified design, but neither of these, called K.2 and K.3, was as good as K.1. In 1787, K.2 was lent to Captain Bligh of the *Bounty* and was taken from him by the mutineers under Fletcher Christian. It was recovered some years later by the captain of an American sealer who, calling at a small island, now named Pitcairn, was astonished to find people there speaking English. He was given

Above: The regulator made by John Shelton in 1769 was taken by Captain Cook on his second and third voyages to be used ashore with a sextant to check marine timepieces and determine the local longitude.

97

the watch and an azimuth compass. The watch came back to England in 1843. K.3 was issued to Captain Cook for his third famous voyage aboard *Discovery*.

PERFECTING THE MARINE TIMEKEEPER

A mechanical timekeeper comprises three main parts. There is the driving force, normally a weight or spring; the time-keeping element itself, usually an oscillator such as a pendulum or a balance wheel and hairspring; and some gearing to keep the oscillator going and to turn the hands.

The oscillator must control the rate at which the spring or weight runs down. Soon after 1700, it was realized by many clockmakers and watchmakers that if there were too intimate a connection between driving force and oscillator, the rate would be controlled not by the oscillator, but by the driving force. John Harrison put it more succinctly: '. . . the Less the Wheels have to do with the Balance, the better'.

The man who first succeeded in designing an escapement to achieve this desirable state of separation between wheels and balance was a French genius of horology, Pierre Le Roy (1717–85). Eventually he

Above : Pierre Le Roy, after a miniature on ivory by Noemi Philastre, was one of France's most distinguished horologists. He was helped, like Harrison, by George Graham.

Right : A nineteenth-century oil painting by Davy of John Arnold and his family. Arnold was the first to patent the spring detent chronometer escapement, in 1787. He made precision pocket watches and longcase regulators as well as marine chronometers.

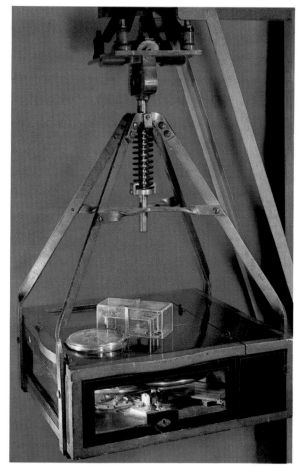

Right : Ferdinand Berthoud's first horloge marine of 1760 which was suspended from a cabin roof. It has two balances and Harrison gridiron compensation. The frictional rest escapement is driven by a spring and fusee, which have to be wound every 24 hours. He made many others, six of which were lost at sea, and introduced a spring detent escapement about the same time as Arnold and Earnshaw.

Right : Pierre Le Roy's marine timekeeper in its gimbals. It was presented to Louis XV in 1766. The large circular balance under the movement had two U-shaped mercury and alcohol thermometers fastened below it. One can just be seen here. The mass of mercury moved at different temperatures to compensate for the temperature errors. The most fundamental invention incorporated was a detached detent escapement.

outshone even his father, Julien Le Roy (1686–1759), also a famed maker, and succeeded him as Horloger du Roi. Henry Sully, the English maker under whom his father had worked in France, gave Pierre Le Roy an introduction to George Graham, who kept him up to date with his horological inventions, setting Le Roy on the path to becoming an innovator himself.

Graham was working on a marine time-keeper which had a cylinder escapement by which he set great store. The cylinder, also called the horizontal escapement, was a marked advance on the verge, but did not leave the balance entirely free to swing because it imposed some friction all the time. Le Roy knew about it and perhaps collaborated in some way with Graham because there are similar drawings among his records.

In 1754, Le Roy deposited at the Académie des Sciences a sealed document setting out his views and a design for a timepiece of novel construction. He referred to Graham's design and the problem of friction, which he said could not be cured by oiling because, on test, a watch with a cylinder escapement gained 12 minutes in 24 hours after being oiled. He rejected all frictional rest escape-ments (as this type is called) and had designed

one that he was going to call an *échappement à détente* (detent escapement).

The detent escapement was the first successful detached escapement. It impulsed the balance and spring in one direction only. As the balance was swinging through mid-position, it unlocked a small lever which released the gears momentarily to give it a push. The gears were then immediately held up by a brake called the detent. Le Roy also provided an extra precaution in the form of a *remontoire*, a small spiral spring, continually wound by the main-spring, which provided constant force to the escapement.

In his marine timekeeper, Le Roy made the balance a single ball weighing around 1 kilogram (2–3 lb.). The timekeeper was designed to go for six hours, because, Le Roy wrote, '. . . there are always people enough on board a ship, who have nothing to do . . . so any of them can easily wind up the clock. . . '.

It was not very successful, but it had the germ of the solution within it. Le Roy made another marine timekeeper in less than two years, then, after a long gap while he experimented, two more, until in 1766 he produced what became the ultimate marine chronometer. In size it was nearer the earlier Harrison timekeepers than the ultimate boxed chronometers, mainly because of the balance, which was of large diameter and had below it Le Roy's temperature compensation, comprising two large alcohol thermometers. He had worked out, some years before Harrison and independently of him, that compensation for temperature changes was best applied directly to the balance.

Thus Le Roy invented the true compensated balance, which he later simplified by using two dissimilar metals for the rim. The rim was made in two semicircular parts, like two scythes, the handles being the spokes, or, as the watchmaker calls them, crossings. The ends of the rim bend inwards as the spokes expand outwards in higher temperatures. Harrison also used a bimetallic strip, but in a different way, to alter the effective length of the hairspring.

Le Roy also discovered that for every hairspring there is a particular length that will give equal times whether the balance has a large or a small arc of swing. Shorter lengths will make the watch go faster when the arcs of swing are large, and greater lengths will make the watch go faster when the arcs are small.

The huge balance wheel of Le Roy's

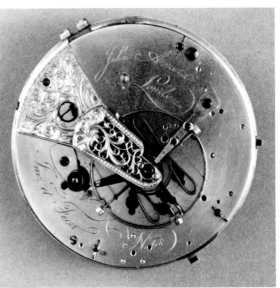

machine was suspended from a harpsichord wire and ran within roller bearings. When the machine, and a duplicate, were entered for the prize offered by the Académie des Sciences in 1766, the wire broke because of the rough ride from Paris to Le Havre in a post-chaise, but Le Roy managed to obtain another locally and repaired the machine on the spot. It turned out that his was the only timekeeper ready for the test on the yacht *Aurore*, which had been commissioned especially for the purpose. The result of three months' trial was very satisfying for Le Roy. His second machine varied no more than seven and a half seconds in 46 days, equivalent to an error of no more than 3 kilometres (2 miles) at the equator.

After a second trial that was equally successful, the Académie awarded Le Roy

Berthoud produced a brilliant array of ideas and a large number of timekeepers, as well as becoming one of the most prolific writers on horology. He may have invented what is called the spring detent, which was incorporated in the ultimate marine chronometer. Berthoud had a habit of exaggerating his own inventiveness and of claiming that others, particularly Le Roy, copied his ideas. In truth, Le Roy's ideas were simpler and more fundamental.

In England, after the final payment to Harrison, Parliament repealed the longitude legislation and introduced another Act offering a prize of £10,000, but the conditions were so onerous that it seemed Parliament was determined that no one else was going to win any money. Nevil Maskelyne, Astronomer Royal, commented that it would 'give the mechanics a bone to pick that would crack their teeth'.

Some of the tooth-cracking is evident in the insults that the professional makers not infrequently threw at each other. The most vituperative seem to have been the chronometer makers, perhaps because the competition was so fierce. Thomas Earnshaw (1749–1829) seldom lost a chance of slanging or poking fun at his rival John Arnold (1736–99), or at anyone who praised Arnold. Both were after the £10,000 prize, and in the process developed Le Roy's chronometer.

Arnold was an incredibly skilful craftsman as well as being an ingenious experimenter and inventor. In 1764 he presented George III with a finger ring in which a watch was set. The watch was about 1.3 centimetres ($\frac{1}{2}$ inch) in diameter, and would not have been unusual except for the fact that it was a repeater – a really complicated half-quarter repeater that sounded the hour, the quarter, and then another blow if more than seven and a half of the minutes of the quarter had elapsed. This gift, and the fact that Arnold could speak German to the Hanoverian King, who had a great interest in horology, helped him to gain the favour of the court and promoted his business.

Immediately after his gift of the watch, he started to develop a marine chronometer with a detent escapement and balance wheel compensated for temperature changes, as a result of which he received some money from the Board of Longitude to help him make two more. They did not perform very well on test. Captain Cook wrote in his journal, '. . . little can be said in favour of the one of Mr. Arnold's on board of us. . .'. Arnold mounted his chronometers direct in

a double prize, one for the timekeeper and the other for his description of it. Some years later the Académie offered another prize, which Le Roy also won. He then seems to have decided that his marine timekeepers were capable of no further improvement. He did not return to his experiments until a few years before his death.

Le Roy spent much of his life and his money working on horological problems, yet his truly original marine timekeeper was developed by the English, not the French, into one that could be made in quantity. For his pains, Le Roy gained a medal.

Le Roy's rival was one of the number of Swiss horologists who found that their inventive capacities were more appreciated outside their own country. Ferdinand Berthoud (1727–1807) went to Paris in 1745, when he was 17, to develop his already considerable skills in watchmaking. He began to concentrate on marine timekeepers in 1762, having become well known in the meantime not only for his ingenuity and excellent craftsmanship, but also for his writings on the subject. This gained him the appointment of Horloger de la Marine, which might have sparked off some of the rivalry in that area with Le Roy.

boxes without an inner case, and on many occasions a spider got in through the winding hole and interfered with the going.

Cook's remark set Arnold on his mettle, and he developed some fine pocket-sized chronometers with helical springs, an improved detent escapement, which still had a pivoted lever to impulse the balance wheel, and temperature-compensated balance wheel. He patented the improvements, including the cylindrical spring, although Robert Hooke really had the priority on the spring. By the last decade of the eighteenth century Arnold had made about 900 chronometers, many of which were doing navigational duty at sea.

Abuse as well as praise was heaped on Arnold, who was well equipped to answer back. His character was an odd mixture of pomposity and braggadocio, more befitting a charlatan than the skilful originator that he was. In 1791 he issued a pamphlet in which he asserted, 'I have been of infinitely more service to my country than any other man'. About his chronometers, he said:

I have lately invented a new mode of escapement of such a nature that friction is utterly excluded from it; and, in consequence, the use of oil, that bane to equality of motion, is rendered wholly unnecessary: and, whether the material be a diamond, steel, brass, or piece of wood, is perfectly indifferent, as they are all equally proper for the purpose.

Thomas Earnshaw could not resist commenting on Arnold's boastful announcement. In a book published in 1806, *Longitude – An Appeal to the Public*, he wrote:

Hear it, ye watchmakers! Friction is utterly excluded from scape and balance pivots; and whether the material be a diamond, steel, brass, or a piece of wood, they are all equally fit for the purpose. Joyful news indeed for the watchmakers, as now they all make wooden timekeepers.

In fact, Harrison had used wood to reduce friction, and diamond was used for the end of a pivot to bear on, also to reduce friction.

Earnshaw had suffered badly from damaging deliberate rumours. He was originally a finisher, responsible for providing the high polish and patterned surfaces traditionally applied to the metal parts of watches. So he finished his earlier chronometers in the same way. His rivals put it about that Earnshaw's chronometers would not work unless a lot of time was spent on finishing them. Earnshaw then began to leave the metal parts unpolished, to prove that the design was sound and did not require high polish to reduce friction. Rival makers then claimed that they put better finishes on their £5 watches than Earnshaw did on his chronometers.

Earnshaw came to London from Ashton-under-Lyne in Lancashire, perhaps soon after 1770, and after working as a finisher became a specialist in jewelling watches and making the tiny cylinders of ruby used in the best cylinder escapements. He made several detent escapements that were pivoted like Arnold's, then hit on the idea of using a short spring instead of a pivot. He showed this to his employer, the chronometer maker John Brockbank, asking him to keep the invention secret. But, according to Earnshaw, Brockbank immediately told Arnold, who patented it only eight days later.

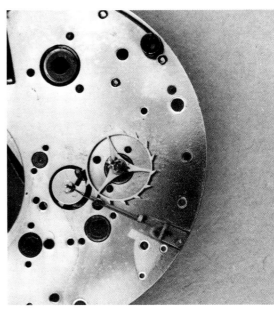

Right : Two views of a pocket chronometer escapement by Robert Roskell of Liverpool and London. The temperature compensation curb can be seen on the left, and the escape wheel and spring detent (the straight piece) on the right, where it is holding up a tooth of the escape wheel. The balance has been removed. Roskell sometimes used the rack lever escapement developed by Peter Litherland of Liverpool.

Having no money to pay the 100 guineas for a patent, Earnshaw went to another London maker, Thomas Wright, who kept the original for a year and then not only patented it (in a different form from Arnold's) in his own name, but had the gall to suggest that Earnshaw make the escapement and sell it to the trade at a guinea over price to pay to Wright to reimburse him for patenting it. He also wanted 'Wright's patent' stamped on every movement!

Earnshaw worked hard, not only at improving his spring detent escapement but also at developing the compensated balance wheel, until both were more satisfactory than Arnold's, and became standard designs for most subsequent makers. But he still had to work for others, and when, after many years, Maskelyne was persuaded to test one of Earnshaw's chronometers, he found the name 'Wm. Hughes' on it, and demanded an explanation. The explanation he received was that Earnshaw had invented many devices but was unable to sell them because others did not understand them. So he was obliged to apply them to watches he supplied to watchmakers for resale, and they put their own names on the watches. In other words, he was a trade supplier without enough capital to set up on his own. Maskelyne not only accepted the explanation but was so impressed with Earnshaw's honesty and ability that he took the trouble to put work Earnshaw's way on several occasions, an unusual gesture for Maskelyne.

In 1791 a trial of chronometers was carried out by Captain Bligh, who was about to set out on a voyage to Tahiti and the West Indies. He was told by the Admiralty

to buy a timekeeper, and of four sent by Arnold, one by Brockbank, and five by Earnshaw for trials at Greenwich, Bligh chose one of Earnshaw's, because not only was the rate better than that of the others, but, at 40 guineas, it was cheaper.

This did not get Earnshaw much further financially. After the trial he sent a petition to the Board of Longitude for money to work on marine timekeepers, but was turned down. A second application caused some doubt among Board members as to whether his claims to certain inventions were valid, but Maskelyne was strongly on his side.

The second application was successful inasmuch as the Board offered Earnshaw £200 for two chronometers. Earnshaw turned down the offer as an insult. The Admiralty came to the rescue by organizing some

Left : Movement of a very high quality pocket chronometer made in London in the nineteenth century by James Ferguson Cole, who was known as 'the English Breguet'.

competitions among makers which Earnshaw won, to gain a number of orders. This encouraged him to make several attempts to win the £10,000 award without 'cracking his teeth'. He did not succeed, although his machines went better than those of any other maker.

Again he petitioned the Board, and this time they agreed that he should have a reward of at least £2500. As soon as this became known, he was fiercely attacked by a number of other chronometer makers for stealing Arnold's ideas. The Board investigated and, according to Lieutenant-Commander Rupert T. Gould, who wrote the history of the marine chronometer, were treated in evidence to a 'good deal of hard swearing' and obscuration of the main issue 'with all manner of envy, hatred, malice and other Christian vices'.

Powerful opposition also came from Arnold's patron, Sir Joseph Banks, who was President of the Royal Society and an *ex officio* member of the Board, backed by the hydrographer of the navy, Alexander Dalrymple. Banks influenced the Board so strongly that they agreed to give Arnold and Earnshaw £3000 each, less what had previously been given them, if they submitted drawings of their chronometers.

Seething under this acceptance of Arnold as on a par with himself, Earnshaw complied with the conditions and received his reward. Arnold was dead when the award was made, but his son received £1678. Earnshaw waited until the specifications were published, and then took an advertisement in *The Times* and another in the *Morning Chronicle*, abusing Sir Joseph Banks and the secretary of the Board. The advertisement was so scandalous that Banks wanted the Board to prosecute Earnshaw for libel.

Typically, the Board took no action, but the hydrographer Dalrymple did, in a pamphlet to which Earnshaw replied in another pamphlet, one sentence of which read, 'In your puffing publication, you appear to have had two motives, one to puff off your darling boy, the present Mr. Arnold, son of your old tutor and friend: another to please Sir Joseph'.

When Arnold died, an obituary in *The Times*, of 17 August 1799, stated:

As a mechanic his abilities and industry will be ever remembered by his country. He was the Inventor of the Expansion Balance, of the present Detached Escapement, and the first artist who ever applied

Left: A two-day box chronometer by Charles Frodsham, London. Two-day chronometers were normally for use at sea. The key, in a hole near the top hinge, was removed and the chronometer turned over in its gimbals for winding through a covered hole in the back. An up-and-down dial showed the number of hours in reserve.

the Gold Cylindrical Spring to the balance of a Timepiece. He retired from business about three years since, but his active mind still labouring for the completion of his favourite object, and for what he called the ultimatum of timepiece making, has produced a Chronometer, far different and infinitely superior to anything yet made public. His son who succeeded him, we understand is in possession of all his father's drawings and models, and from him we may now hope for the completion of that grand object – the discovery of the Longitude by Timekeepers.

Although his rival was dead, this was too much for Earnshaw to stomach. He was sure that Arnold's son had written the obituary. Of it he wrote:

There is no truth in this publication, except that of his father living at Well Hall, dying on Sunday morning, his having retired from business about three years, and his son succeeding him. If he was the first who applied gold springs to watches, it was because the corrosive matter which (unfortunately for him) always ouzed from his hands, rusted all the steel ones. The truth of this all his workmen knew. . . .

VI
THE DEVELOPMENT OF THE WATCH

Egocentric man must have become aware at a very early stage in his history that his shadow accompanied him wherever he went in the sunshine, and that it changed its length as the sun went down.

The length of a man's shadow changes considerably during the day. Even in the Mediterranean basin it can be as long as 7.3 metres (24 feet) in the morning or evening and as short as 1.2 metres (4 feet) at midday. At some significant moment in history, a man thought of measuring the length of his shadow. He did so by placing a stone on the ground and moving towards the sun until the tip of his shadow fell upon the stone. Then he placed the heel of one foot to the toes of the other and measured the number of feet to the length of his shadow, by counting them on his fingers.

This is not an imaginative supposition. Several references are made in older writings to measuring time by stepping it out along a shadow. They are, however, much later than some known shadow clocks that still survive. The Greeks commonly arranged meetings by the feet in a shadow, and Aristophanes (c. 448–c. 388 BC) mentions in one of his comedies that meal times were also fixed by the length of shadows.

To measure the time in feet is not as crude as it may seem. The assumption is made, with some measure of validity, that the taller a man, the longer his foot, so that the proportion stays relatively constant. In the fourth century AD, the Roman Rutilius Palladius wrote a treatise on agriculture entitled *Opus Agriculturae*, in which he gave the time in feet for various months of the year.

One of the earliest personal timekeepers was what the Egyptians called a *merkhet* and the Greeks a *horologus*. Its primary use was as a holder for a plumb-line, but it could also check the transit of selected stars across the meridian to calculate the hour of the night because the position of a 'clock star' in the sky at night will provide the time like the position of the sun in the day. There is one of these in the Science Museum, London. According to its inscription, the museum example was the property of Bes, an astronomer-priest of the god Horus of Edfu in Upper Egypt. The bronze of which it is made is inlaid with electrum, a natural gold-silver alloy.

The most common personal sundial was a cylinder, perhaps of clay, with different hour scales around it for different months.

Left : The watch held by Helena Leonora du Booys- de Sieveri in this portrait by C. Jonson van Ceulen has an enamelled case showing one of the favourite designs of the Huaud family, the Geneva enamellers.

Left : A gilt and metal enamel verge watch. The movement is signed Quare, London. The subject and composition was commonly used by the Huaud family, as can be seen in the portrait opposite.

Left : The meridian can be determined by looking due south (in the northern hemisphere) across a plumb line, and the time at night by when certain stars cross it. That was how the Egyptians used this portable merkhet, c. 600 BC, made of bronze inlaid with electrum, a gold-silver alloy.

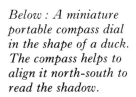

Below : A miniature portable compass dial in the shape of a duck. The compass helps to align it north-south to read the shadow.

It was hung from a thread and a horizontal pin was inserted at the top of the appropriate month scale. The tip of the shadow down the cylinder gave the time. The Romans carried such sundials and the Saxons apparently developed a similar system or copied the idea. A tablet made of silver, hung on a gold chain, was found in the soil of the Cloister Garth at Canterbury Cathedral in 1939. It is marked in tides on each side for the months, and the pin or gnomon is a gold pin with an animal's head set with jewelled eyes at one end. When not in use the pin is stored in a hole in the bottom of the tablet. The tablet is inscribed in Latin.

A favourite form of compass dial in seventeenth-century Europe was a tablet folding in two, bookwise. The tablets were made in silver, ivory, boxwood or brass. When the halves were opened, a string stretched across them as the hypotenuse of a triangle. The shadow of the thread showed the hour when the dial was placed on a north-south line indicated by a small compass needle in the base. Some had series of holes through which the string could be threaded for different latitudes.

The first portable mechanical timekeeper was made possible by the invention of spring propulsion. It then became simply a matter of mechanical ingenuity to reduce the size sufficiently for the timepiece to be carried on the person, in other words, to become what is today called a watch. The solution was found in one of those simple ideas that, once thought of, seem so obvious. The early clock frame has a plate top and bottom, kept apart by corner posts. The pivots of the wheels run in vertical pillars.

Right : Model of a tenth-century Saxon pocket sundial made of silver with a gold cap and chain, which was found in the soil of the Cloister Garth at Canterbury Cathedral in 1939. The gnomon is a pin with a jewelled end in the form of an animal's head. It shows the time by the length of its shadow when the sundial is suspended. The pin fits into a hole in the base when not in use.

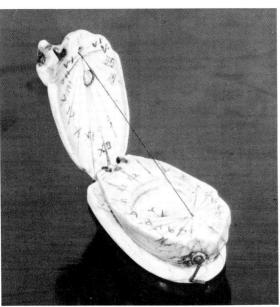

Musk-apple watches struck the hours, and so they should properly be called clock-watches. But they were really hardly more than a novelty. They were about 8 centimetres (3 inches) in diameter, and had a small-diameter pedestal foot with a thick rim at the bottom. This was the dial on which the hour was shown, by a short, stubby hand. One of the first centres of manufacture was Nuremberg, and some years ago the watches were wrongly named 'Nuremberg eggs' through mistranslation of a passage in Rabelais: Rabelais referred to little living clocks (*ueurlein*), and the translator mistook this for *eierlein*, or little living eggs.

The main production at Nuremberg soon became the drum-shaped watch, often with a striking movement. The movement was of

Someone thought of turning the wheels sideways so that their pivots ran in the plates. The layout was simpler and more logical, but required more precision in manufacture. The first known illustration of this layout appears in a Burgundian or Flemish manuscript, *L'Horloge de Sapience* (The Clock of Wisdom), written between about 1460 and 1480.

EARLY WATCHES

Clocks with plated movements were first mounted in drum- or canister-shaped cases with the dial on the top, and were placed on a table. Later they appeared in square cases. The earliest watches discovered, however, are in ball-shaped cases. They were made in the first half of the sixteenth century and it is surmised that this kind of watch was invented about 1500. The reason for the ball cases is a combination of logic and fashion. At the time, when sanitation was crude, pomanders or scent bottles containing musk were worn on a ribbon or chain around the neck. The bottle was a hollow metal sphere with pierced decoration, and an ingenious clockmaker thought of putting a spring-driven movement in a 'musk apple', as they were called.

iron and the gilt case was at first like a round box with a separate lid. Later the case had a hinged lid that was decorated and pierced to reveal the numerals on the dial. There was no glass at this time. The single stubby steel hand showed the time on a 24-hour dial, but it rotated once in 12 hours, so the dial was marked from I to XII and the numbers were repeated in Arabic, 13 to 24. Knobs at the hours enabled the time to be felt in the dark.

Towards the end of the sixteenth century, the box case was disappearing and the hinged-lid type was beginning to become more sophisticated, with rounded sides and domed top. The French were now leading the way in design and innovation. To overcome the varying power of the mainspring, the Germans introduced two devices. One

Above : One of the first clocks for wearing on the person, a 'musk apple' watch by Jacques de la Garde of Blois, with its fusee striking movement. The earliest watch still in existence is by the same maker and is dated 1551. (It is illustrated on page 57.) The dial and single hand are on the foot.

Left : A late Renaissance travelling alarm clock made in Lübeck by Nicolaus Siebenhaer. The small knobs on the outer double-XII chapter ring are touch knobs for finding the time in the dark. Inside is a 1–24 ring of Arabic numerals, then a double-XII ring. There is a moon dial in the zone.

Right : An enamelled gold watch (c. 1650), made before the invention of the hairspring by Estienne Hubert, of Rouen, with a portrait inside the cover and one hand.

simply allowed the spring to operate over only a part of its unwinding. Watchmakers call such a mechanism stopwork.

The other device was a cam turned by the spring. A small wheel on the end of a powerful, short spring pressed on the edge of the cam. The idea was to smooth the power output by extra friction, which decreased as the spring ran down. This is known as a stackfreed, from the German *starken feder* ('strong spring'). It was never satisfactory.

As the watch was carried on the person, it was natural for the jeweller or goldsmith to take on the task of casemaking. An associated craft was boxmaking, which introduced the watchmaker to the already flourishing craft of the enameller. The most important early centres were Limoges and Blois in France and Geneva in Switzerland. Holland was another key centre and Flanders, London and Paris also had their enamellers.

At first, cases were made in the traditional way of the enameller, by engraving or carving hollows in the metal to take the molten coloured glasses (*champlevé*), or by soldering on strips to form a pattern of compartments for the enamel (*cloisonné*). Then, in about 1630, Jean Toutin invented a way of painting pictures in colour without

having to run the enamel into separate chambers, which not only considerably simplified the task of the casemaker, but gave the enameller much greater freedom of design. Cases of the period are often covered with flowers, although sometimes a scene or a portrait was the theme. The Huaud family of Geneva became particularly famous for their enamelled cases.

Early watches had a drilled knob on the side of the case through which a ring passed, at right angles to the watch dial, to take the ribbon or cord without twisting. This was later replaced by a loop parallel to the dial. The key for winding was separate, and was usually attached to the same chain as the watch to avoid its being lost. Early French watches with one hand, known as *oignons* because of their thickness, were usually wound through the centre of the hand.

The watch was far from accurate, but it was pretty, so it was worn more as jewellery than for timekeeping. The timekeeping element was an oscillating bar with knobs on the ends called a dumb-bell balance, or it was a wheel balance. The technical breakthrough came with the application of a spiral spring to the balance wheel. Suddenly the watch became reasonably accurate.

Right : In the centre is a French chatelaine with a verge watch by Vauchez, Paris, and beside it on the left a Swiss miniature watch with visible balance dating from about 1820, and a Swiss ring watch (c. 1800) with a musical movement on the right. At the bottom on the left, the automata watch (c. 1790) shows a tightrope walker performing, and on the right, a clock-watch by Daniel de St Leu, London, made for the Turkish market.

*Right : At the top is an
alarm watch in a silver
pair case by Thomas
Smoult, Lancaster,
(c. 1710), and below
it, two wandering hour
watches, an English
one on the left, signed
Antram, London, and
a German version on
the right, signed
Gottfried Torborch,
Munchen, (c. 1710).
In the centre is a watch
with an experimental
dial (although it
cannot be seen as the
case is closed), this is a
sun and moon watch by
Joseph Buckingham,
London, c. 1700.
Lower, left and right,
are two fat French
oignon watches,
(c. 1700) both alarms,
by Clouzier, Paris, and
Jean Rousseau, and at
the bottom a third
oignon without alarm
by Chas. Champion,
of about the same date.*

Above : A German sixteenth-century watch and the movement of another, both with stackfreed control. The stackfreed – the curved spring with a roller on the end that presses on the edge of a cam on the mainspring arbor – can be seen on the left. The toothless part of the toothed winding wheel is the stopwork.

The invention of the spiral hairspring by Huygens in 1675 coincided with a fashion for men of wearing a waistcoat. Instead of being hung round the neck or from the belt, a watch was dropped into a pocket in the waistcoat and was kept safe by attaching a chain to the loop at the top called the pendant. When the watch disappeared into the waistcoat pocket there was not so much occasion for showing it off, so the case became plainer, and also thinner for comfort. In England Puritanism was another influence which encouraged plainness. However, decorative enamelled watches were still much in favour with the ladies, who wore them on chatelaines.

One influence of the hairspring was that it encouraged the rapid introduction of a minute as well as an hour hand. Although

clocks with two hands had been in use for about 25 years, this method of indicating time was still unfamiliar to many people. There were protests from time to time about adding another hand to a clock or watch to confuse the owner.

This inspired watchmakers to invent alternatives to concentric hands. During the last quarter of the seventeenth century some returned to one of the oldest ideas of all, a revolving dial. It was a version of the Italian night clock. There is a semicircular opening in the dial round which an hour numeral moves, its position in the semicircle indicating what portion of the hour has elapsed. There is no hand.

Another version of the same idea, called the sun and moon dial, has a similar semi-circular aperture marked in 12 hours from 6 to 12 and back to 6 again. During the day, an image of the sun moves round the hour scale, and during the night, an image of the moon. Minutes are shown by a hand in the normal way.

Some watches had dials marked in six hours from I to VI, and underneath each hour chapter another six hours from VII to XII. This meant that the single hand went twice as fast, and minutes could be marked by large divisions. Some, like the regulator clock, had separate hands with separate dial rings. Today they seem more confusing than concentric hands.

Because of a general knowledge of the accuracy the pendulum gave to clocks, some makers thought they could sell watches more effectively if their customers thought they were pendulum-controlled. A few, not understanding that the operation of the

Left : During Puritan times in England, watchmakers produced 'Puritan watches' which were small and plain. This one is by Richard Crayle.

Right : The first drawing of a balance spring fitted to a watch, as shown by Huygens in the Journal des Scavans. *It was made by Isaac Thuret in Paris, on 22 January 1675, for Huygens. Thuret claimed the invention, but later denied his claim.*

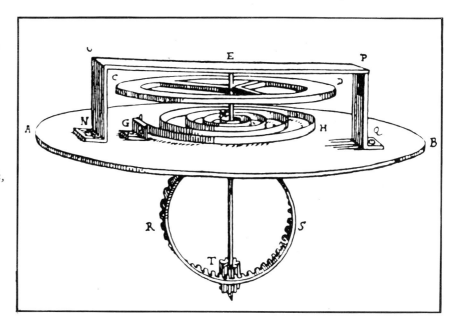

Above : An enamel pendant watch in a rock crystal case made in the second half of the seventeenth century by Hans Otto Halleicher, Augsburg. A single hand shows the hours and there are age of moon and calendar dials.

TOMPION AND HOOKE

The most famous watchmaker of this period was Thomas Tompion (1639–1713), who worked in Fleet Street. He was a clockmaker as well, as was the custom of the time, although the skills requisite for making a watch were becoming more and more divorced from those involved in making a clock. In his lifetime his workshop produced about 6000 watches and about 550 clocks.

When he died it was written of him:

> Mr. Tompion, of London, one of the most eminent persons for making clocks and watches that have been produced in the last age, dyed last week. Indeed he was the most famous and most skilful person at this art in the whole world, and first of all brought watches to anything of perfection . . . He was acquainted with the famous Dr. Hooke, grew rich, and lived to a great age. He had a strange working head, and was well seen in mathematicks.

Tompion was an exceptionally clever and versatile craftsman who understood mechanisms and could grasp fresh ideas, but he was not inventive in a general way himself. His initial success was the result of an introduction to Hooke when he came to London from Bedfordshire, via Buckinghamshire, where he seems to have worked for a while. Tompion was then about 39, and both men were bachelors, although their private lives were very different.

Tompion was a model of rectitude. In contrast, Robert Hooke (1635–1703) led a bohemian life. His lodgings in Gresham College, Bishopsgate Street, where he was professor of geometry and where the Royal Society met, tended to be full of squabbling women. A niece, Grace Hooke, came to live with him when she was 11, and he lavished care and gifts on her. When she grew up she became his mistress, but she left him to get married. He continued to pursue her, none the less. He also had a girl called Nell Young to live with him for a time until she too departed to marry someone else. Hooke also chased after her.

The Royal Society asked Hooke to get a quadrant made for no more than £10. He seems to have been dissatisfied with the Society's instrument-maker, as he waited until he could find another craftsman, who turned out to be Tompion. The quadrant was a success, and Hooke commended Tompion at some length to members of the Society. After that they met frequently, often at one or another of the coffee houses that

pendulum depended upon the pull of gravity, even believed that a pendulum would work in a watch. They mounted a little pendulum on the back instead of the balance wheel. There is in existence one watch with a mini-pendulum between the plates, like a clock.

Most pendulum watches made at around the end of the seventeenth century are, however, novelties. A small disc, usually engraved with a circle, was fastened to the rim of the balance wheel, and the metal bridge over the wheel was pierced so that the disc could be seen swinging to and fro like the mock pendulum on the dial of some bracket clocks of the period. A few watches were made with the disc swinging in a slot in the dial. Others had a glass in the back of the case so that the 'pendulum' could be seen.

Right : A sun and moon dial is a feature of this late seventeenth-century watch by Peter Garon, London. The moon shows the hour in the semi-circular aperture at night and the sun by day. The large hand indicates minutes and the small one at the bottom, seconds. The time shown is just after 02.49 a.m.

abounded in old London.

Hooke was a top-level craftsman himself, but his chief attribute was his inventive mind, and he spent much time helping Tompion with ideas. In return, Tompion closely co-operated with Hooke in everything he wanted made. During a bitter argument which Hooke carried on with Oldenburg (the Royal Society's secretary) and Huygens, about the priority of the balance spring for watches which Hooke also claimed to have invented, Tompion worked feverishly with Hooke, sometimes 12 hours a day and for weeks on end, making demonstration watches to show the King, from whom Hooke was trying to obtain a patent. He took Tompion with him on two occasions, which no doubt helped with Tompion's royal commissions.

Unlike some members of the Royal Society,

Hooke treated craftsmen as equals. However, his friendship did not stop him from chiding Tompion from time to time for slowness. 'Tompion a Slug', reads one of the notes in his diary.

The watch that was offered to the King as evidence did not get Hooke a patent, to his chagrin. The cover of the balance was hopefully inscribed 'R. HOOK *invenit an.* 1658 T. Tompion *fecit* 1675'. This cover later came into the possession of George Graham, Tompion's journeyman who became his friend, married his niece, and succeeded to his business. Graham was reported as saying that Tompion made parts for Hooke at the early date, and Hooke would not disclose their purpose. Tompion, it seems, certainly believed Hooke to be the originator of the balance spring.

Left : A pendulum watch of the end of the seventeenth century. The 'pendulum bob' is a disc attached to the balance and seen in the top curved aperture. Below it, the hours are shown in digital form through another aperture.

Right : A verge escapement watch movement made by Thomas Tompion and Edward Banger, his apprentice. Later, Tompion dismissed Banger and did not refer to him again.

The watches of Tompion and his contemporaries represent the first period of relative accuracy, when a minute hand was added. There was little improvement in the escapement or the balance and spring, except that the spring became longer. The escapement remained the verge.

NEW INVENTIONS

In the meantime, other escapements were invented. One, the cylinder escapement, developed in about 1725 by George Graham, after pioneering efforts by Thomas Tompion, became more popular on the Continent than in England after Julien Le Roy in France enthused over one sent to him by Graham. By a twist of fortune, the duplex escapement, invented by Le Roy's son Pierre, became popular some years later for the better English watches. The escapement in general use remained the verge, however, and precision watches of the later eighteenth century incorporated the detent escapement.

What did improve watches considerably was an invention of quite another sort. It was not an English invention, but it was patented and developed in England, which gave English makers an advantage over their rivals. Even with a spiral spring, a balance running in metal bearings was erratic. Running in jewel holes, the friction was reduced and the timekeeping was thus much improved.

The first to propose using drilled gemstones as bearings was a Swiss named Nicholas Facio (1664–1753), who moved to England in 1687. He was a mathematician, astronomer, religious enthusiast and political intriguer, and became a protégé of Isaac

Right : Side view of a George Graham watch movement with a cylinder or horizontal escapement, showing the boot-shaped teeth at the top.

Left : An engraving of George Graham by T. Poyley, published in 1820, based on an original portrait.

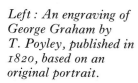

from the watchmakers, but also from the jewellers, diamond cutters and engravers of precious stones. In their petition against the Bill, the watchmakers made the pained complaint, 'though we retain our compassion towards the persecuted Protestants of *France*, we could never expect that they would attempt to take away our Livelihood . . . These three forward *French-men* can't pretend that an Act of Parliament, made here, can keep the pretended invention in England. . . . '

The Worshipful Company of Clockmakers produced as evidence a watch made by Ignatius Huggeford with a large jewel set in the centre of the decorated cock over the balance wheel which pre-dated the patent. As a result, the Bill was rejected. The watch was of so much use to the Committee of the Clockmakers Company that the records for 15 January 1705 instructed the Renter Warden to buy it, to be kept as evidence. This he did, for £2 10s.

The watch is still in the Company's collection, and the extraordinary fact is that when it was re-examined in the last century, it was found that the jewel is merely a decoration and was never intended as a working jewel, despite its position. Whether

Right : The back of the watch (c. 1675) by Ignatius Huggeford, London, which was offered in evidence by the Clockmakers Company of early jewelling, to oppose an application for the extension of the first watch jewel bearing patent of 1704. In fact, the jewel in the centre of the circular cock was a decoration and not a bearing.

Newton. He was elected a Fellow of the Royal Society at the age of 24, only a year after moving to England. One of his projects was for the religious conversion of the world, and another was for raising the dead. With two French watchmakers, Peter and Jacob de Baufre, Facio obtained a patent for watches with jewelled bearings in England in 1704, and advertised in the *London Gazette* that jewelled watches could be obtained from their shop in Church Street, Soho.

In a statement issued later to back an application for a Bill in Parliament to confirm the patent, they pointed out that as the world was overstocked with watches, the trade in England was 'very languishing', and their invention would revive it for twenty to thirty years.

This raised a storm of protest, not only

or not the witnesses were aware at the time that the jewelling was fraudulent we shall never know.

Henry Sully said that in 1704 he had seen Sir Isaac Newton with a watch that had a diamond with a hole drilled through it as a bearing. It was Peter de Baufre's club-footed verge. Facio and the de Baufre brothers managed to keep their methods of drilling jewels a trade secret for a long time. It did not become known on the Continent until about the 1770s, which was to the disadvantage of continental makers. Earnshaw complained that he had been asked for 100 guineas just for a look at some jewel-making tools. Ferdinand Berthoud, the great French horologist, was sent to England as a kind of industrial spy to discover the secrets of John Harrison's marine timekeeper. In a letter dated 14 March 1766, he wrote, 'If some parts of this watch would be difficult to make, there are others which could not be done at all in France. I mean the pierced rubies carrying the pivotal staffs'.

As late as 6 September 1771, Pierre Le Roy confirmed this with the words, 'Whereas in Harrison's watch and indeed in all the good watches of England, the balance-staff and the last wheels are set and move in pierced rubies, we in France have not the secret of making these rubies'.

The only maker outside England who used pierced jewels was Breguet, some of whose early *perpetuelle* (self-winding) watches had English jewelling. His friend John Arnold probably supplied them, as a letter written later by Arnold refers to a delivery of jewels. At the beginning of the nineteenth century, Breguet had his own jeweller to make them, a man named Hooker, whom he had brought from England to Paris 'at great danger and expense'.

About this time, the French began to make their own jewels, followed by the Swiss, but still only for top-quality watches. It was not until a later maker of jewels for the firm of Breguet, Pierre Frédéric Ingold, returned to his home in La Chaux-de-Fonds, Switzerland, and set up a jewel-making factory there, in 1823, that watchmakers began to jewel everyday watches.

Jewels known as endstones were fitted by English watchmakers before the introduction of pierced gemstones, and afterwards. They are still used today. The function of an endstone is to reduce friction when the pivot is vertical and turning on one end. Endstones were first applied to the fast oscillating balance staff, which carries the balance and

Above: A Charles Goode watch in the British Museum. It has one jewel only, a diamond endstone in the balance cock. The diamond is not facetted, as were most.

spring. At the back of the watch is a metal piece called the balance cock, elaborately pierced and decorated in earlier watches, which holds the bearing hole. A diamond or ruby with a flat surface was often fitted over the hole for the staff to pivot on while the watch was lying dial up.

Diamonds had been brought into London from India for a long time, and these were supplemented substantially from around 1730, when Portuguese merchants began bringing rough diamonds into London from the new fields in Brazil. These were rose-cut for jewellery, as were Indian stones, and this was also suitable for jewelling watches. After about 1750 many better-grade watches had both pivots of the balance running in jewel holes and endstones as well.

Two further important technical improvements were made to watches in the early eighteenth century, the first by the English watchmaker Henry Sully (1680–1728), who worked in France. He had the idea of cutting little cups, called oil sinks, around pivot holes to stop oil from being drawn away, or spreading away, from the pivots.

Sully was of such remarkable character that he is worth a short aside. Julien Le Roy named him 'horology's martyr', because he

gave so much to horology and received so little from it. As a youth he was apprenticed to a respected London maker, Charles Gretton, and perhaps through him or Daniel Quare, whom he also knew, met Christopher Wren and Isaac Newton. He obviously impressed them, because apparently they encouraged him in the idea he had for a timekeeper to work at sea.

He must have been quick to grasp ideas because he was soon as much at home conversationally with scientists as he was with clockmakers and watchmakers. According to Le Roy, Sully went to Holland as a young man and learned Dutch and French, then on to Vienna, where he quickly learned German.

In Vienna he was patronized by Prince Eugene, the Duke d'Aremberg, and Count Bonneval, who were serving with the army on the Rhine and appointed him their horologist. When Austria made peace with France, the Duke d'Aremberg took Sully to settle in Paris and provided him with an income.

In Paris Sully met the Maréchal de Noailles and John Law, the Scot who had become the French Minister of Finance, and later was responsible for the 'Mississippi

Above: A French watch by J. Jaque André Bosset (c. 1700) with an enamelled cock showing a girl with a canary. A small silver heart attached to the balance swings over her head.

Right: A late eighteenth-century quarter repeater watch with diamond endstone by Eardley Norton, in an 18-carat gold case. The dust cap is shown below and the outer case on the right.

ESCAPEMENTS

Here are four escapements used successfully
in watches and also, to a lesser extent, in clocks.
In each case a circular balance and spring is
centred on the smaller circular part. The
toothed wheel in each case is the escape wheel.
Right: The cylinder escapement, also called
the horizontal escapement, which has friction
rest and is more compact than the verge and
crown wheel which it replaced in watches.
Centre right: The duplex, with separate teeth
for impulse and locking. It has friction rest
and was first used for more expensive and
later for low priced watches. Below left: The
final Swiss form of the jewelled-detached
lever escapement which superseded all other
forms of escapement for the watch. Bottom:
The detached chronometer or detent
escapement, in this case a spring detent, used
in the nineteenth century for precision watches
and marine chronometers. It is occasionally
used today, but note that the Swiss call
precision lever watches 'chronometers'.

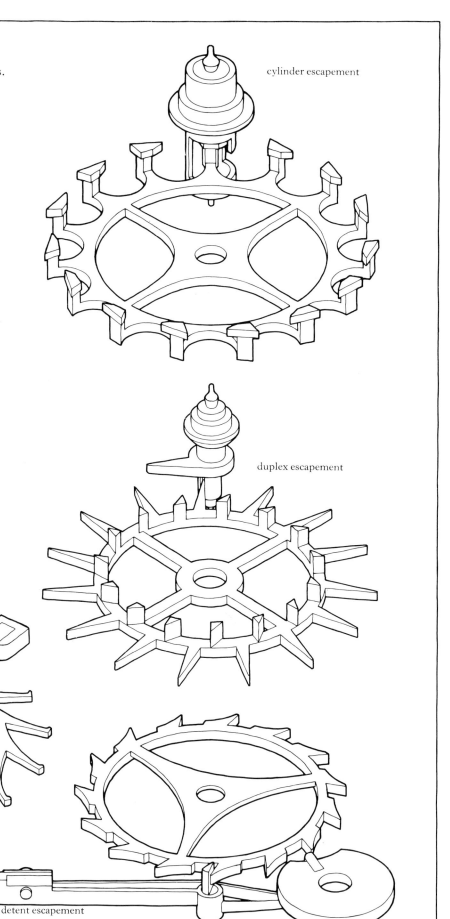

cylinder escapement

duplex escapement

detached lever escapement

detent escapement

bubble', the financial disaster in the French colony of Louisiana. In an effort to resuscitate French watchmaking, these two provided the financial backing to enable Sully to set up a factory in Versailles and then one in St Germain.

Sully persuaded 60 watchmakers from London to join the Versailles venture, but it was not successful, although it had considerable influence on the industry. The second factory failed because of a financial depression. A scheme was then proposed and accepted to transfer Sully and his men to London, where, with the backing of the Secretary of State, he thrived, but when the Secretary died, life became difficult for Sully and his French wife and they had to return to Paris. There, too, his rich and famous patrons ignored him and he had to work as a repairer.

Nevertheless, he completed a marine timekeeper which won him a pension from Louis XV. Alas, while he was testing it, his tools and furniture were seized because he had been unable to pay his rent. He fell ill, never recovered, and is buried at St Sulpice. It was left to Sully's friend Julien Le Roy to perfect the oil sinks.

The other invention was made by one of the patentees of watch jewels, Peter de Baufre (fl. c. 1689–1722). He invented a frictional rest escapement that was deadbeat (that is, did not recoil). It was taken up in England, where it became known as the club-footed verge, and also in France, with modifications by Sully and by Jean Paul Garnier. About a century later, it was revived for machine-made watches in Lancashire in another variation and became known as the chaff-cutter or Ormskirk escapement.

REPEATING WATCHES

Repeating watches were very much in favour in the upper end of the market, because they sounded the time in the dark, a valuable feature when artificial lighting was by lamp and candle.

Rack striking, invented in England in 1676, prevented the striking from becoming out of step with the hands, and so was particularly useful for clock-watches (watches that strike the hour), because a watch, wound daily, was much more likely to become out of time with its striking than a clock wound weekly. Also, it was more of a nuisance to open a watch and adjust it.

However, it was not for regular striking that the rack was introduced, but for its

ability to strike the last hour or quarter on demand. Two makers claimed the invention of a quarter repeater. The Reverend Edward Barlow (1636–1716) made a watch with two push pieces, one for repeating the hour and the other for repeating the last quarter – a ting-ting for every quarter of an hour passed since the hour. Daniel Quare (1649–1724) made one with a single push piece that caused the hour and then the quarter to sound, and won a patent for it from James II in 1687, in competition with Tompion, who used Barlow's system.

The notes were struck on a bell in the back of the case. In a quarter repeater, a single note indicated the hour and a double one the quarter. The half-quarter repeater was also popular. This sounded a triple sequence on a single bell. The first single notes indicated the last hour, the next double notes the quarter, and a final single note was struck if another seven and a half minutes had passed since the quarter, thus ting, ting, ting . . . ting-ting, ting-ting . . . ting indicates a time between 03.37½ and 03.45.

The more elaborate minute repeater took longer to become established, probably because it cost more. Again a single bell was used in the same way, except that the last

Above : The facetted diamond endstone, known as a 'rose' in the diamond trade, became common in higher quality watches of the second half of the eighteenth century.

notes, struck rapidly, indicated the minutes passed after the last quarter, so that ting, ting, ting, ting . . . ting-ting . . . ting, ting, ting indicated 04.18.

For the buyer who could not afford the minute repeater, but wanted something superior to the quarter repeater, there was the five-minute repeater. The system was again similar, except that every single note in the final sequence indicated that another five minutes had passed after the last quarter.

Minute repeaters, now much sought after by collectors, were produced only to special order until about 1800, but after that some were made for general sale.

Later repeaters were sometimes fitted with a pulse piece. When this was pressed, the bell (or gong in later watches) was silenced and, instead, the hammer tapped the end of the pulse piece so that the time could be felt, which was useful to a deaf watch-owner.

In the eighteenth century, both Julien Le Roy and Abraham-Louis Breguet slimmed their repeating watches considerably by eliminating the bell and using a steel rod or rods as gongs. Both also used a steel block for some watches, called dumb repeaters because the blows on the block could be felt, rather than heard.

ENGLISH AND SWISS COMPETITION

The English and the Swiss were the most successful watchmaking nations for most of the eighteenth century. By about 1750 the English were making most of their higher-quality models with jewelled bearings. The Swiss fitted hardened steel plates, called *coquerets*, instead of endstones.

The Swiss had come on the watchmaking scene about the same time as the English, in 1600, when Huguenot watchmakers from France and Germany swelled the numbers of those working in the two countries. The Swiss specialized in watches of various shapes, called form watches. Some of the cases were made of brass and many were metal or rock crystal. Shapes included cruciform; faceted ovals; skulls; books; birds or animals; and tulips and other flowers.

From 1601 Geneva had statutes regulating the activities of watchmakers, and these were revised in 1673 to allow only citizens and burghers of Geneva to become masters. There were provisions against competition from other towns and from outside Switzerland and also against women, who for a quarter of a century were only allowed to

Top left: Thomas Earnshaw made this half-quarter repeating watch movement. Working in High Holborn, London, he produced repeating, cylinder, duplex, lever, and chronometer watches.

Above: A large musical watch by T. Williamson, London. The watch is in a pair case and has an enamelled dial. Two tunes are played on six bells, the hammers being operated by a pin-barrel.

Left : A Geneva enamelled gold watch set with half pearls. Engraving in little scoops under transparent enamel gives an effect called guilloché.

Below : Three Geneva enamelled watches, the cruciform dates from the later nineteenth century, the centre watch is by the Frères Dubois after 1850, and the form watch on the right after 1800.

make fusee chains, the tiny chains used to equalize the pull of the mainspring.

While England was in a turmoil of invention, Geneva was building a big industry and doing so more rapidly than other countries, by concentrating on production. Genevese watchmakers did very well in exporting, although, as Zedler's *Universal Lexicon* of 1746 relates:

> The English watches are considered best of all, especially the so-called repeating watches. After the English watch come the French, Augsburg, Nünberg and Ulm watches. The Geneva watches are thought little of, because they are to be had so cheaply; they are made in such quantities that one buys them in lots.

The Swiss adopted an invention of the French horologist Jean Antoine Lépine (1720–1814), who abandoned the fusee and used a series of bars to hold the pivots of the various wheels, instead of a plate. The Lépine calibre, as it was called, made the movement much thinner because the balance wheel was between the plates and bridges and the going barrel (a spring barrel that is rewound in the same direction as it drives) was thinner than the fusee.

Right : Jean Antoine Lépine introduced his Lépine calibre about 1760 in which separate bars were used instead of a top plate. The idea was developed in such movements as this, to make them thinner and easier to service.

Left : A cruciform watch by Charles Bobinet, Paris, c. 1640. The mainspring and fusee are across the arm of the T and the balance is on the stem.

By the beginning of the nineteenth century, about 50,000 watches a year were coming from Genevese workshops, about 12,000 of them in elaborate enamelled cases, and about 6000 repeaters. The special feature of Geneva enamels was an overglaze. After the colourful enamelled picture had been painted on the case, an almost transparent glaze (pale yellow for gold and copper and pale blue for silver) was laid over the top and machine-polished to give the appearance of transparent glass.

Novelty watches continued to be big business for the Swiss in the later nineteenth century. Many were repeaters, the hour just passed being sounded on a gong. Most of these repeated the quarters as well; some were minute repeaters. Musical watches were made in large numbers in Geneva, and

automata on the dials were popular, showing a little figure striking a gong, Father Time striking with his scythe, a woman at a spinning wheel, children see-sawing, a mill wheel and such like. For special customers there were pornographic watches with animated bedroom and rustic scenes. These were not always on the dial; often they were under a cover, and some had secret releases for the lid.

BREGUET

One name in watchmaking outshines even those of Tompion and Le Roy and represents another era of watch development. It is that of Abraham-Louis Breguet (1747–1823). Breguet was Swiss, but went to live in France after his father died and his mother married his father's cousin, who was a watchmaker working in France. Young Breguet took to watchmaking at an early age, and when he was 28 he married a French girl whose dowry enabled him to set up in business at 39 Quai de l'Horloge, in l'Ile de la Cité, by the Pont-Neuf in Paris, which was the horological and instrument-making quarter at the time. Breguets still live there.

The novelty that first made Breguet famous was his *montre perpetuelle*, a self-winding watch wound by a bouncing weight, in the same way as a pedometer works. The second one, made in October 1782, was supplied to Queen Marie Antoinette. It repeated the time on gongs on demand and also showed the date. The Queen bought several Breguet watches, even ordering one while she was in prison. She never paid for it, according to Breguet's account books.

Breguet changed the watch from a rather

Above : An engraved portrait of Abraham-Louis Breguet by A. Chazal.

better'. In his *tourbillon* watches he made the entire escapement revolve, thus eliminating certain timekeeping errors, and he also calculated a special shape for the hairspring which eliminated more timekeeping errors. This is now known as the Breguet overcoil.

Breguet was also concerned with blind owners and made special watches called *à tact*, meaning that the time was ascertained by touch. Around the case were knobs representing the hours. On the back or front cover was a large hand that could be turned until it was resisted by the watch. Its position relative to the touch pieces gave the time. Twelve was, of course, indicated by the pendant. Such watches were also useful for sighted people in the dark, and for noting the time unobtrusively when the watch was in a pocket.

Breguet watches often break auction price records today. In Breguet's own time so many forgers flourished by using his name that he had to take special steps to combat them. He provided his watches with a secret signature on the dial, below the number 12. This signature is so fine and small that on a silver dial it can only be seen if the watch is held at a certain angle in the light; on enamel dials the signature is only visible under a strong lens.

It is said that between 1790 and 1823 500 fake Breguet watches were exported from Paris to every genuine one. The fakes were often typical fat watches of the time in pair cases, and not at all like Breguets. When a well-known collection of watches was put on exhibition in New York a few years ago, two of the three Breguets turned

bulbous timekeeper in a separate outer pair case to a much slimmer pocket watch. His craftsmanship was superb, and he contributed several fundamental inventions, one of which was the shock-absorber, which he called his *parachute*. He demonstrated it by dropping his watch on the ground. When it did not stop, Talleyrand is said to have remarked, 'This devil Breguet is always trying to go one

Right : The dial and back of a complicated carriage watch by Staples, London, c. 1790. It shows the time, day, date, and sidereal time, strikes the hours and quarters, and plays tunes on six bells. On the back a human figure in three parts keeps changing and below it some 'curtains' can be drawn aside to reveal an animated pornographic scene.

Left : An early nineteenth-century repeating automata watch made in Paris shows two figures striking on two bells.

Below : A repeater watch with automata made in the first quarter of the nineteenth century by Meuron and Co., Paris.

accurate watches to be sold to the public – pocket chronometers with detent escapements.

Arnold and Breguet very much admired one another's work. Arnold sent his son, John Roger Arnold, to work with Breguet for a time and, after Arnold had died, Breguet sent John Roger an Arnold pocket chronometer that he had converted to a *tourbillon*, as a token of esteem. It was engraved 'Breguet's Homage to the revered memory of Arnold'.

There were two types of detent escapement, the earlier being pivoted, and used mainly on the Continent, and the other mounted on a spring, and used mainly in England. Arnold invented the spring detent about 1780, and Breguet was one of the few continental makers to adopt it. A

out to be fakes made at the time of Breguet.

Breguet's customers included Napoleon and Josephine. Napoleon took a Breguet watch on his Egyptian campaign but sand got into its movement. On his return he demanded a new one in exchange. No one argued with Napoleon; he got it. The Tsar of Russia would carry only Breguet watches, and after Napoleon's death King Louis XVIII showered honours on the watchmaker. Breguet watches are carried by the characters of Alexander Dumas, Stendhal, Balzac, Pushkin and many other novelists. According to Jules Verne, the watch Phineas Fogg took with him in *Around the World in Eighty Days* was a Breguet.

THE PRECISION WATCH

Breguet worked in the period of the precision watch, which can be regarded as dating from about 1750 to 1830. It was sparked off by the intense effort in all maritime countries to conquer the navigation problem. John Harrison showed that a high-precision watch was possible, even with a verge escapement. Pierre Le Roy provided the means and John Arnold developed the methods which led to the production of the first truly spring in place of the pivot forestalled the problem of oiling; when the oil thickened, the detent did not move so freely. Arnold also developed the temperature-compensated balance to the form it ultimately reached.

Many fine pocket chronometers were made by Edward John Dent (1790–1853), who was in partnership with John Roger Arnold for ten years, then set up nearby in opposition and eventually won the contract to make the Big Ben clock. Victor Kullberg (1824–90), a Swede who moved to England in 1851 and worked in London, was another maker of fine pocket and marine chronometers.

Detent escapements for chronometers had to be filed individually by highly skilled specialist detent makers, so production of chronometer watches was limited. For their

Right: A gold quarter repeating watch by Breguet with double parachute shockproofing. It also shows the date, moon phase and power reserve.

Below: A gold cased bras en l'air *watch, c. 1810. Pressing the pendant moves the arms to indicate the time.*

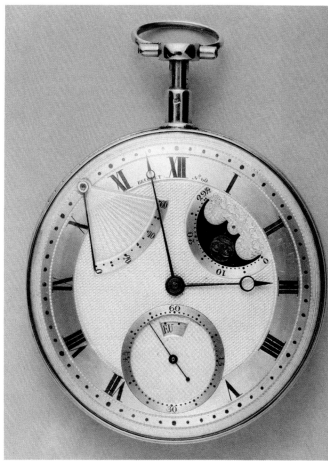

second-quality watches, English makers adopted the duplex escapement, which has two sets of teeth on the escape wheel, at right angles to each other. One set impulses the balance wheel, and the other locks the gear train between impulses. It is a frictional rest escapement, like the verge and the cylinder, and unlike the chronometer, which is detached.

The duplex was favoured by some top makers, including James McCabe and David and William Morice, and continued to be made until the beginning of the twentieth century, even though the detached lever escapement had by that time become almost universal.

THE LEVER ESCAPEMENT
Breguet used the cylinder or horizontal escapement in his earliest watches and throughout his working life. For some watches, in order to reduce friction, he improved a ruby cylinder escapement made by English jewellers. The English ruby cylinder followed much the same pattern as the usual steel cylinder of the time, with a pivot at each end and the escape wheel teeth running between them. It was a comparatively fragile arrangement because

WATCH OSCILLATORS AND HAND-SETTING

On the right is the balance and spring, the most successful oscillator for 300 years for portable clocks and watches. Centre right, an electro-mechanical impulsing system applied to a balance reacts with magnets below it through a mechanical switch. The next step (below right) was the tuning fork (1960) which was controlled by a transistor and drove the hands mechanically. The final step was to use a strip of quartz that oscillated and drove the display electronically. Left below are the forms of clutch used in watches (a friction pipe or tube) and clocks (a spring washer) to enable hands to be turned when setting them.

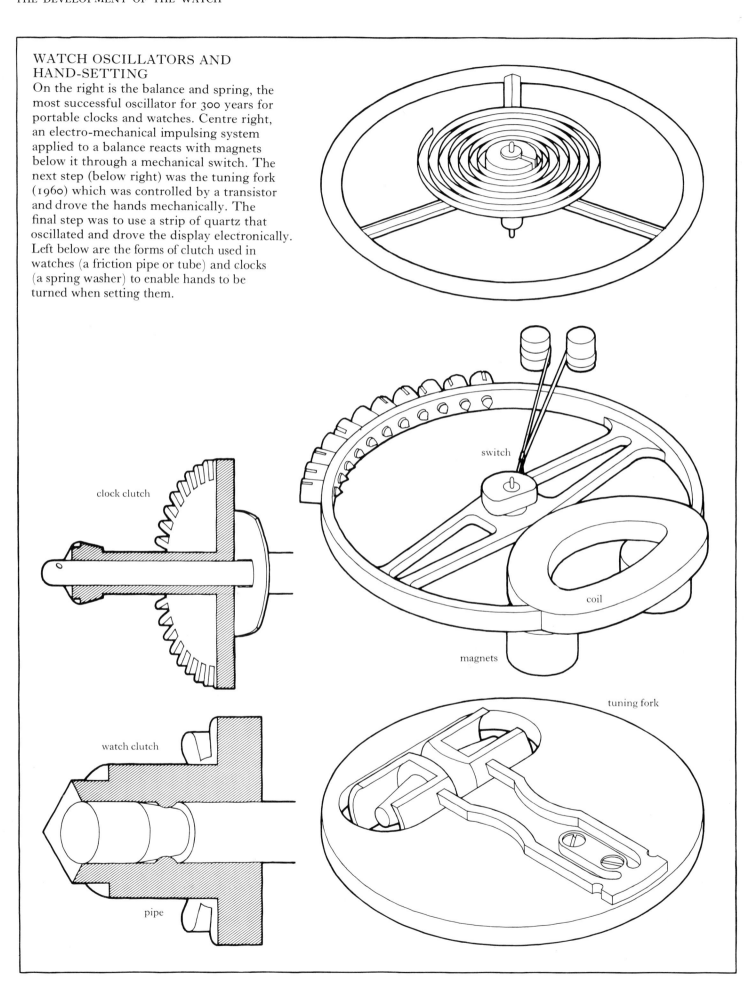

clock clutch

watch clutch

pipe

switch

coil

magnets

tuning fork

the ruby half-cylinder took all the loading. Breguet made the cylinder staff solid, running in two pivots, with the cylinder, and escape wheel, above the top pivot.

He did not favour the duplex escapement because it did not perform so well in his thinner watches, but he did appreciate the value of the lever escapement, which he first made in 1786 and continued to fit to some of his watches until he died. He developed the detached lever escapement to the point where it differed only in detail from the millions of lever escapements used in mechanical watches and clocks today.

The inventor of the lever escapement was Thomas Mudge (1715–94), the son of a Devon clergyman. He was apprenticed to George Graham in London, an excellent start for a young man. Mudge might even have been on the premises when John Harrison first called on Graham, as he would then have been 15 years old. After Graham died, in 1751, Mudge set up on his own in Fleet Street, but even before then he had acquired such a reputation for superb craftsmanship that the King of Spain gave him *carte blanche* to select and supply any watches or clocks that he cared to send.

Above: The oscillating weight which bounces to wind a Breguet montre perpetuelle. *The pendant is at the top when the watch is in a pocket so that the pedometer weight can bounce as the wearer moves.*

Right: A gold Breguet souscription *watch with ruby cylinder escapement and parachute shock protection. The dial has a single hand and secret signature.*

Far right: A Breguet silver and gold mounted tourbillon *watch, in which the escapement revolves every four minutes. It is engraved 'Pour le Prince Repnin en Novembre 1810'.*

Mudge was one of the committee appointed to report on John Harrison's H.4 timekeeper, and, as already related, was accused by Harrison of having disclosed its secrets to the French. Mudge had been cautious enough, before joining the committee, to publish a tract giving his own 'Thoughts on the means of Improving Watches and More Particularly Those for the Use of the Sea', so as to avoid being accused of stealing Harrison's ideas.

According to his patron, Count von Bruhl (an avid follower of horology and astronomy), Mudge began working, in about 1754, on a watch escapement that was similar to the anchor escapement which had been so successful for pendulum clocks. The lever was shaped something like an anchor and in fact is called the 'ancor' on the continent. At the end of the 'hook' part are pallets that engage alternately with the teeth of the escape wheel. The stem of the lever has a slot at its end. Near the centre of the balance and spring is an impulse pin which enters the slot momentarily in mid-swing of the balance. When this happens, a tooth of the escape wheel is unlocked which moves the lever smartly to the opposite side, impulsing the balance through the impulse pin. The escape wheel is held up again by a tooth being stopped by the opposite pellet. The process is then repeated in the opposite direction as the balance swings back.

The first lever watch made by Mudge was either bought by George III and presented to Queen Charlotte or presented to her by Count von Bruhl, and it became known as 'the Queen's watch'. It now belongs to Queen Elizabeth and is kept at Windsor.

The case of the Queen's watch is of 22-carat gold and is 5.7 centimetres (2¼ inches) in diameter. It is surprisingly thin. The watch stands for display on a contemporary plinth finished in gold and tortoiseshell, which has been modified to hold it. Because the lever escapement was the biggest single invention in the field of watchmaking after the balance spring, this watch has been described as the most important in the world.

The gold case bears the hallmark for 1770, so perhaps von Bruhl was referring to models or drawings that Mudge may have made in 1754. A model made by Mudge is known. He made at least two bracket clocks with lever escapements, one in about 1759, which is much earlier than the watch. It is small for a bracket clock of the time, and has

Above: A Breguet montre à tact *with a cylinder escapement in a gold case. Turning the hand until it stopped indicated the time in the dark or to a blind person.*

Left: A parachute shock absorber (top left) of the Breguet type, here used in an unsigned French watch in a 14-carat gold case of about 1810.

Right : Thomas Mudge, in an oil painting by Nathaniel Dance.

Below : Queen Charlotte's lever watch, kept in Windsor Castle, which is the first ever made by Thomas Mudge with his lever escapement. The case is hallmarked for 1769.

elaborate moon work. It once belonged to the famous engineer Isambard Kingdom Brunel, whose grandmother was Thomas Mudge's sister-in-law, and it is now in the Ilbert Collection at the British Museum.

There is also a travelling clock with a jewelled lever escapement by Mudge that is thought to pre-date the Queen's watch. It is drum-shaped, with an 18 centimetre (7 inch) round dial, strikes the hours and half-hours, and repeats the hour when a cord at the bottom is pulled. The clock belongs to Lord Polworth, one of whose ancestors married von Bruhl's daughter.

Count von Bruhl urged Mudge to make another lever watch, but he pleaded that he was too busy with his marine timekeepers, although he did provide the Count with a lever escapement model in 1782.

However, another early lever watch came to light in Devonshire in 1960, belonging to Major Eric Flint, a direct descendant of Mudge. It was once an exact replica of the Queen's watch and had a lever escapement. Unfortunately, it has been converted to spring detent escapement and the compensation has been removed. Around the dial is inscribed, 'The original was invented and made by Thomas Mudge for Her Brittanic Majesty A.D. 1770'. The case hallmark is for 1795, after Mudge's death, and the watch was probably made by his partner, William Dutton, who is known to have made another such watch in 1800.

After his examination of Harrison's H.4, Thomas Mudge became so obsessed with conquering the longitude problem by making a perfect timekeeper that, in 1771, he moved to Plymouth to devote himself to this, and to winning the awards offered under the parliamentary legislation (and also to be near his brother, Dr John Mudge, a physician and maker of reflecting telescopes).

He finished his first marine timekeeper in 1774, which could not have been more badly timed, for it was the year of the new Act of Parliament, reducing the prize to £10,000 and making the conditions almost impossible to meet. Even after this, Mudge held the opinion, quite wrongly, that a prize under the previous Act would be open to him.

The timekeeper followed the same principles as Harrison's H.4, with a form of verge escapement and a *remontoire* to make sure that a constant force was supplied to it, temperature-compensated balance springs (Mudge used two, Harrison one), and maintaining power to keep the watch going while being wound.

The escapement was more complex than Harrison's; the crown wheel wound up two small spiral springs (the *remontoire*) alternately, which impulsed the balance. It was an idea that could have benefited precision pendulums; both Riefler and Strasser later invented successful escapements for observa-

tory clocks which worked by loading a constant force spring to impulse the pendulum. The temperature compensation in the Mudge marine timekeeper was the same as he had used in the Queen's watch.

Mudge lavished skill on this timekeeper: it has been described as the most beautifully finished watch ever made (although, in fact, when it was examined in recent times, it was found that the pillars had not been very neatly riveted and the plates were only partly gilded). It is mounted in an octagonal gilt case with glass panels, and has four enamel dials (hours, minutes, seconds and reserve power), surrounded by filigree work.

When submitted to Greenwich Observatory for testing, it was carried daily across a courtyard, which caused it to stop after three months. It failed a month later with a

Below : Parts of the Mudge lever watch. The balance and the cross-shaped balance cock (mounted on corks for photography) are at bottom centre. The movement was jewelled with rubies for the balance, sapphires for the train wheels, and milky sapphire for the pallets and notch stones.

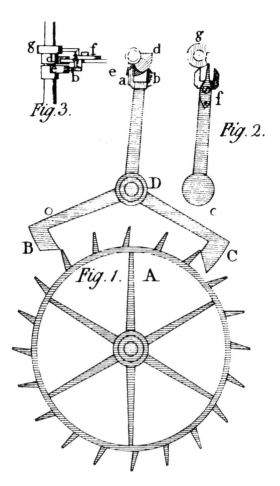

Left : The first lever escapement used in the Queen's watch, from a drawing in a book by Mudge's son on his father's timekeepers, published in London in 1799. It is reminiscent of a dead-beat clock escapement.

Below : The first clock with a lever escapement, made by Thomas Mudge in about 1760. The large circular balance is supported on anti-friction rollers.

and wrote subsequently that he never intended to make money out of it, 'as long as I am capable of amusing myself with it, it will serve for a hobby-horse, and when I can no longer do that, I will destroy it'. He never did. Von Bruhl bought it and it is in private hands today.

The two other timekeepers were alike and went for 36 hours. Each had two enamel dials set in filigree work. They were called 'the Blue' and 'the Green' after the colours of their cases, and were beautifully finished despite the fact that Mudge's sight was failing and he had to work to a large extent by touch. They did not perform well at Greenwich and, eventually, the Board rejected them.

By the time of the last trials, from 1789–90, Mudge was suffering from senile decay, and

broken mainspring. After complaining to the Astronomer Royal, Nevil Maskelyne, about this treatment, Mudge submitted the timekeeper again two years later, in 1776, when it was kept in one place. After testing, Maskelyne reported to the Board of Longitude, 'it is greatly superior, in point of accuracy, to any timekeeper which hath come under his inspection'. After he had said that he wanted to make two more marine timekeepers, Mudge was awarded £500.

The first machine was kept on test, but the mainspring broke again. Eventually, it was abandoned for service as unsatisfactory. Mudge had designed it to run for eight days, whereas Harrison's H.4 had to be wound daily. To avoid the strain on the mainspring which caused the trouble, Mudge fitted a stop that limited the going to two days. Much later it became general practice for sea-going chronometers to run for two days and those for depots to run for eight days.

Like Harrison, Mudge became totally disenchanted with the Astronomer Royal and the Board. He let Count von Bruhl have the machine, and apparently it performed excellently on voyages with Captain Campbell. He never submitted it again to Greenwich,

Left: After being appointed a member of the committee to examine Harrison's No. 4, Mudge became more and more interested in marine timepieces. This is the movement of his first (which he always referred to as his 'watch' because it did not strike). Later he made two copies, known as the 'Blue' and the 'Green'.

his lawyer son, Thomas Mudge of Lincoln's Inn, applied to the Board, unsuccessfully, for a reward. He then prepared a parliamentary petition and published a pamphlet violently attacking Maskelyne. After receiving a committee report favourable to Mudge, Parliament voted £2500 more to him, despite the strongest opposition from the Board. Not much later, in 1794, Mudge died.

Ironically, the lever escapement had been ignored by him in his search for precision timekeeping at sea. In a letter he wrote to von Bruhl, he said, 'And as to the honour of the invention, I must confess, I am not at all solicitous about it: whoever would rob me of it does me honour'. However, in the same letter he pointed out that the lever escapement would out-perform, in a pocket watch particularly, any other escapement known

at that time; but 'you will find very few artists, and fewer still that will give themselves the trouble to arrive at; which takes much from its merit. . .'.

Mudge was correct in his supposition. Von Bruhl lent Mudge's model to a Swiss watchmaker working in London, Josiah Emery, with the request that he make a lever watch. Emery was reluctant to accept the commission, but, eventually, he completed his first lever watch in 1795 and he continued to make about three a year, which he sold for 150 guineas each.

With the detached lever escapement, the balance is disconnected from the rest of the movement while swinging, except for a few thousandths of a second when it receives its impulse. What Mudge did not fully appreciate was that his lever was not safely 'locked' in position when not moving. It was held by friction alone and a jerk would dislodge it. The solution was to alter the geometry so that the pallets, which were impulsed by the teeth of the escape wheel, were drawn into them when not in operation.

Draw was first introduced by John Leroux in the watches made by him in London from about 1785. They are the only known eighteenth-century watches with draw.

Even Breguet did not at first appreciate the value of draw, and he did not incorporate it in his watches until about 1814. Breguet's first escape wheels had pointed teeth, like those of the English watches, except that there was a little flat on the ends to combat wear. The point of a tooth moved a pallet on the end of the lever. He then made the tips of the teeth wider, so that the work of 'lifting' the lever (impulsing the balance) was divided between the tooth and the pallet. This divided lift became the feature of the modern club-toothed Swiss lever. Breguet also invented the jewelled pallet in its modern form, as well as what are called the horns, dart and roller, which prevent the escapement from malfunction if the watch is knocked. It took him 26 years to introduce these changes, but they have all proved fundamental.

The first progress towards universal use of the lever, was, strangely, a backwards step. Peter Litherland, a Liverpool maker, took out a patent in 1781 for a rack lever escapement, in which the lever swung the balance to and fro through the intervention of a rack on the lever and a toothed pinion on the balance staff. The balance lost all the advantage of being detached.

Nevertheless, the idea lent itself to production in quantity, and Lancashire rack levers performed quite adequately when well made. The Abbé de Hautefeuille had invented a similar escapement, but with a coil spring, about 1676. Litherland probably never knew about this. Rack levers were made until about 1830, when Liverpool makers began to use detached levers, based on Breguet's designs.

Although Mudge's lever watch for the Queen was well known at the time, it is possible that the lever escapement was invented quite separately in France. There are some who think Breguet may have invented it, but this is unlikely because he did not claim the invention (and he was never backward in that respect), and he was in close touch with what went on in England. However, Julien Le Roy may well have invented a form of lever escapement.

A Swiss collector, Dr E. Gschwind, found a movement in a Paris street market in the 1950s and bought it for a few francs. The balance wheel is inset with 22-carat gold studs to increase the weight of the rim. The escape wheel has 12 teeth and the anchor has steel rollers (instead of jewels) as pallets to engage the escape wheel teeth.

The lever seems to be original, since there are no signs of conversion from verge escapement, and it is thought that it may have been made by Le Roy. However, this is by no means certain, as the movement cannot be dated with any accuracy. The kind of maintaining power based on Harrison's model and introduced to France by Berthoud in 1768 was once fitted. It is known that Le Roy used an anchor in 1755 for slowing down the striking of a repeating watch (the idea is still used today for operating alarm bell hammers).

Below : Mudge's copy of his first marine timepiece which is known as the 'Green', after the colour of its case. It was once thought to have been lost at sea.

Above : Three lever watch movements. Bottom right is a rack lever by Robert Roskell, Liverpool ; on the left, one by Yeates and Sons with Massey's crank roller lever ; and above, a watch with pump winding (pressing the pendant winds it) by Adam Burdess, Coventry, 1869.

Right : A Liverpool watch movement with a rack lever escapement and diamond endstone. The semi-circular scale is for the regulator.

Le Roy and Mudge knew each other, and Mudge certainly conveyed technical information to Berthoud. If Le Roy did pre-date Mudge, one would have expected something to have been written about it at the time, but nothing was, by Le Roy or anyone else.

Le Roy was another maker who devoted his later life to trying to 'solve the longitude', so perhaps, like Mudge, he abandoned his lever escapement.

THE KARRUSEL

Bahne Bonnicksen was a Dane who settled in England and eventually became a British subject in 1910. He instructed students of the British Horological Institute in Clerkenwell, London, for a time, but settled in Coventry, a watchmaking centre then, where he took out a patent for a watch he had been working on

since about 1890. He called the watch a *karrusel*, from the Danish for roundabout, because the escapement revolved as it did in Breguet's *tourbillon*. The difference was that the *karrusel* was much more robust and the rotation occurred every $52\frac{1}{2}$ minutes, instead of once a minute as in Breguet's watches.

For several reasons, the rate of a watch depends upon its position. One is that the centre of gravity of the balance and spring changes, another that the friction of the pivots changes. A pocket watch is normally kept in an upright position in the pocket, although it is taken out and held horizontally for short intervals when it is consulted. A watch may gain a certain amount of time, say, if the pendant is to the left, and lose a different amount when the pendant is to the right, while, with the pendant upwards, the rate is different again. The idea of the revolving escapement is to take it through all these positions to average out the differing rates.

Bonnicksen's *karrusel* watches were successful for many years in gaining high rating certificates for timekeeping at the National Physical Laboratory trials at Kew in Surrey (these were called 'Kew certificates'). Some other makers were allowed by licence to fit *karrusels*, one being Hector Golay, a Swiss

who imported Swiss *ébauches* (unfinished watch movements) and finished them at a workshop in Spencer Street, Clerkenwell, London. Golay supplied unfinished movements with *karrusels* to the watchmaking firm of Rotherhams of Coventry. Rotherhams were bombed out in the Second World War, but they saved a small stock of watches with Golay movements which were sold off.

Bonnicksen was inventive in other directions. He was one of the first to develop a speedometer. A present-day speedometer is a simple analogue instrument that shows a rate of change (miles or kilometres in an hour) instantaneously. Bonnicksen used a clock to provide the time and an odometer (mileage recorder) to provide the distance. These operated a mechanical calculator to show the number of miles per hour.

Above : Julien Le Roy (1686–1759) celebrated French clockmaker who produced thin repeating watches and also devised the horizontal arrangement for turret clocks, first used in England for 'Big Ben'.

Left : An unsigned watch (c. 1810) with a Bonnicksen karrusel revolving escapement.

138

VII
MASS PRODUCTION

Although early clockmakers and watchmakers in the North American colonies were from Europe, and worked with the methods and in the styles they had learned during their days of apprenticeship in England, Holland, Germany, Sweden and elsewhere, in America the trade began to draw in young men who rejected the ideas established over centuries. Their early faltering steps towards producing clocks from the easiest available material, wood, eventually transformed clocks all over Europe.

EARLY AMERICAN CLOCKS

The more wealthy seventeenth-century settlers brought clocks with them, first lantern clocks, then longcase clocks. Turret clocks, because their weight and size made them difficult to transport from Europe, and because they were not for personal use, were rare in the American colonies and were among the first clocks to be made there.

Just before the dawn of the eighteenth century, there were definitely makers of longcase clocks in the Philadelphia area, in the Quaker colony of Pennsylvania, where many wealthy merchants lived. Almost all worked in the English idiom.

Perhaps the earliest clockmaker in Philadelphia was Samuel Bispham (*fl. c.* 1696), although the earliest longcase clock surviving, dated 1709, was made by Abel Cottey (1655–1711). Another early maker was the Quaker Peter Stretch (1670–1746). All of these were English. Because fine furniture was in demand, cabinet makers flourished, and some of them specialized in clock cases when the industry got into its stride around 1750.

The growth of new communities, and sometimes increasing competition, sent clockmakers and watchmakers into New Jersey and New York, where a few isolated makers had set up workshops even before 1700, and on into Delaware, Maryland and Virginia. They spread to Ohio in the west and still farther on into Indiana.

One of America's most celebrated clockmakers of the later eighteenth century was

Far left : An American mantel clock with an eglomisé panel in the door. The maker was perhaps E. C. Brewster. The style was popular around the 1860s.

Left : The pin barrel, bells and hammers of the magnificent musical clock by David Rittenhouse, the Philadelphia clockmaker, shown on page 141. It plays ten tunes (four at any one time) on 15 bells. The wheels at the top operate a planetarium in the dial arch. The clock was made in 1774 for $640.

Left : A tall clock made by Aaron Willard about 1810 in Boston. It has a round painted dial with Arabic numerals, and a pierced fret to the hood. The height is 2.53m (8 ft 3 ins).

David Rittenhouse (1732–96), who first built up a business at Norriton, now Norristown, and later moved to Philadelphia. He was also a surveyor, making his own instruments, and was responsible for surveying two state boundaries, one being that between Pennsylvania and New York. After about 20 years he moved to Philadelphia, where he established America's first observatory. Even being a pioneer astronomer was not enough for this gifted man, although he suffered constant illness. He also became director of the United States Mint, and succeeded Benjamin Franklin, who was his friend, as president of the American Philosophical Society.

His finest works include astronomical clocks, usually with orreries showing the moving planets. When Rittenhouse involved himself in politics, Thomas Jefferson tried to deflect him back to his clockmaking skills, writing, in 1778, 'you should consider that the world has but one Rittenhouse, and that it never had one before'.

The Philadelphia School is considered to have been the best of the early centres. It was in decline in the 1830s and eventually died around the 1850s, when swamped by machine-made clocks from Connecticut.

It was at the rival school, based in Boston, that the first immigrant positively known to have been a clockmaker, William Davis, set up in business in 1683. Bostonian clocks were also very English in style.

This industry became scattered over the states of Massachusetts, Connecticut, New Hampshire, Rhode Island and Vermont. In many areas, clocks were made down to a price for relatively poor rural customers.

One of four clockmaking brothers of the Boston school deserves a special niche in American horological history; for he was the first to break away from the traditional long-case clock by making the forerunner of what became the typical American shelf clock.

The family was named Willard and the idea for the Massachusetts shelf clocks they began to make might have been suggested by the small grandfather clock – the grandmother clock – which was made in some numbers in America.

The first shelf clocks, made around 1800, were shaped like small chests standing on yet another small chest, both of mahogany and about 1.2 metres (4 feet) or so high overall. Other makers copied the idea with variations until the machine age was responsible for standardizing the shape, about half a century later, into the shallow box form now regarded as typical.

Left : The frontispiece of the song 'Grandfather's Clock' that was so popular it gave the tall clock or longcase clock a new name which has persisted until today. The song was written in 1876.

Right : The clock known as David Rittenhouse's masterpiece. It shows the positions of six planets, the moon's orbit, the positions of the sun and moon in the zodiac, the equation of time and a number of other indications as well as playing ten tunes on bells.

Benjamin, Simon, Ephraim and Aaron Willard were four of eight sons of a clock-making father, and they and their progeny had an influence that spread throughout New England as more and more makers reduced their dependence on the longcase clock in favour of the Massachusetts shelf clock, which could give a good account of itself in timekeeping because of its quite long pendulum, form of anchor escapement and weight drive. The rectangular weights hung each side of the case and drove through a system of pulleys.

It was also a rival to the other commonly made clock, the 'wag on the wall', a weight-driven clock with short pendulum that had to be wound daily against the weekly winding necessary for the shelf clock.

It seems that Benjamin and Ephraim concerned themselves mainly with longcase clocks, while Simon and Aaron became more and more interested in experimenting with wall and shelf models. It was Simon who became the most famous of the brothers. One of his designs, which he called, at the time of the patent, his 'improved clock', is now known as the banjo clock, and is a type much sought after by collectors today. The banjo clock is very American in its design.

Left: A wall clock in a two-tier mahogany case by Aaron Willard of Boston.

Right: Simon Willard's lighthouse clock, made about 1800, which was intended as an alarm, although a few were made without it. The name 'lighthouse' was applied later.

Centre right: A shelf clock by Aaron Willard, which is 77cm (2 ft 6 ins) high. Aaron and his brother Simon produced many wall and shelf clocks, particularly after 1800.

Far right: A banjo clock made by Aaron Willard junior in 1841. He often used wooden panels in later versions of the banjo clock because eglomisé painted glass was becoming harder and harder to obtain.

His patent covered the shape of the case as well as technical details of the movement. The front panels of the banjo-shaped wooden case were of glass, decorated on the back with geometrical patterns of the type the French call *eglomisé*, which became universal for shelf clocks. The glasses with geometrical patterns were probably supplied by two ornamental painters known to have been working in Boston at the time. William Fish, a Boston cabinet maker, provided the cases, which varied in finish, the earliest having simple banded inlay along the front edges and an acorn finial at the top. Later, an eagle became a favourite finial.

Simon Willard gave up making longcase clocks when his banjo clocks became firmly established favourites. Many other clockmakers copied Willard's designs, presumably

some of them by licence from Willard, but after the patent ran out, there was a spate of copies bearing the words 'Willard's patent' and even, in some cases, his name painted on the glass in the trunk. These fraudulent practices were not unusual at the time.

This stung Willard into taking a newspaper advertisement in August 1822, six years after the patent's expiry, to caution the public against 'the frequent impositions practised, in vending spurious Timepieces'. Declaring that they were palmed upon the public by the statements on them, he added that several brought to him for repair 'would certainly put the greatest bungler to the blush. Such is the country inundated with, and such, I consider prejudicial to my reputation; I therefore disclaim being the manufacturer of such vile performances'.

Above: A clockmaker's workshop c. 1740, in a print by Martin Engelbrecht, Augsburg. The frame of a watchman's clock and a foliot are propped up against the anvil base in a way reminiscent of the Stradanus print on page 62. Again, the bench work, such an essential part, is very much in the background.

Right: An engraving showing Henri de Vic, supposedly the maker of France's first clock, in the fourteenth century.

Another of Simon Willard's inventions was not so successful. Now called a lighthouse clock, owing to its wooden base and glass shade on top, resembling an early lighthouse, it was in fact designed as a special alarm. The hammer rapped the top of the wooden case instead of hitting a bell. Willard believed that a noise like someone rapping on the door was a more natural way of being woken up than the sounding of a bell. Few buyers seemed to agree with him.

Willard also made tower clocks to order. One which he made for the United States Senate in Washington was lost when the British set fire to the city in 1814, but in 1837, when he was aged 83, he received a commission for two more clocks from Washington, one for the Senate chamber and the other for Statuary Hall. He retired at 85 and lived for another two years.

The story of clockmaking in America is the story of many ingenious men with small businesses, suffering from shortage of materials, problems of distribution and of collecting debts in expanding communities, with the inevitable difficulties involved in keeping afloat in inflationary times. The War of Independence affected almost everyone, as did the 1812–14 war with Britain. The severe restrictions on imports hit clockmakers very hard, as they had depended on brass and brass movements and on painted dials from England.

Probably these dangerous, exciting and formative times fuelled the drive towards self-sufficiency. Whatever the reason, it was then that the root methods of mass production were established, mainly by trial and error, with many spectacular failures and bankruptcies before the general principles were understood and could be applied economically and effectively.

The clockmakers were in the van of mass production. The first domestic article to be successfully mass-produced was the long-case clock, and the pioneer was Eli Terry, an American clockmaker.

FIRST STEPS TO MASS PRODUCTION

A traditional craftsman making muskets, clocks, or anything else, fabricates the entire article, suiting each part to another. In early days, he would make his own screws with threads that were most convenient to his workshop. Each wheel would be made to gear into a particular pinion, and would gear into another only by chance. There was no point in working to an arbitrary standard.

Some early manufacturing centres had the same kinds of pressures that are generated by expanding industry today – how to find enough skilled craftsmen, how to improve methods to keep costs down and maintain or better performance, where to find alternative raw materials when costs become uneconomic, how to expand production while cutting prices, and how to stay in business against others struggling with the same problems.

One of the first steps towards mass production was the division of labour, the benefits of which had already been discovered in Nuremburg and Augsburg in the sixteenth century, when specialists made decorated strip for use as the sides of table clocks. The medieval method of one craftsman, one article, was vanishing slowly as articles became more complex.

Its end was hastened by the growth of craft guilds in the sixteenth and seventeenth centuries and earlier, which laid down standards of training, workmanship and quality, enforced them rigidly, and controlled the entry into particular trades. It then became difficult for craftsmen to be skilled in and to practise several crafts.

In this era, the maker of a lantern clock would probably buy rough castings for the

Above : Part of the American section of a collection of over 400 clocks belonging to Henry Schottler. Here is a variety of shelf clocks and a banjo wall clock.

Left : An old clockmaker using a bow and the turns, which produce more accurate work than the lathe as the workpiece is held between dead centres. In the factory age, however, it was soon superseded by the lathe.

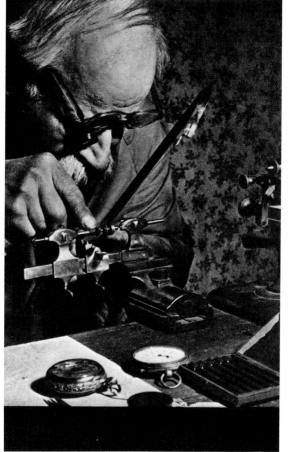

Above : A reconstruction of Samuel Deakin's workshop of 1771 at the Newarkes Houses Museum, Leicester. The forge and anvil are in evidence, as in many earlier prints. The wheel tooth cutting engine in the foreground is a forerunner of production tools.

Right : A large engine for cutting clock teeth, the correct number of teeth being obtained by the holes around the dividing wheel. The illustration was made in 1850.

posts of the movement, the doors and the dial, from a brass-founder, and the bell, perhaps, from a bell-founder. He would file and finish the brass, and probably make the iron parts himself. A watchmaker would depend upon a jeweller for a gem-set case, and later, in England, would buy pierced jewel bearings from the specialist in London. In the first part of the eighteenth century, there were multitudes of specialist suppliers of parts to clockmakers, many of whom became finishers and assemblers.

The maker of clocks and watches from the very early days had one essential tool, called, as it still is, the turns. Turns are a primitive form of lathe, operated by a bow, as they had been in ancient Syria from 4000 years BC.

The tool changed little over the years, except for being made in metal. The piece to

be turned into a round rod is mounted between two fixed points and the string of the bow given a twist around it. Moving the bow, usually held in the left hand, backwards and forwards spins the workpiece first in one direction and then in the other. A graver with a sharp cutting edge is held in the other hand against a rest, and shaves the workpiece.

Another version of the turns for larger work was the pole lathe. The principle was the same, but, instead of a bow, there was a springy wooden pole above the turns, which were mounted on a bench. A cord attached to the free end of the pole passed round the workpiece to be turned and was attached at the other end to a treadle. Pole lathes were still in use in the workshop of A. T. Oliver at 25 Spencer Street, Clerkenwell, London, where he made gold pocket-watch cases by traditional methods, until the premises were compulsorily demolished in 1970.

The lathe turning in one direction only was invented in the fourteenth century, as far as is known. The workpiece was still rotated between fixed centres, but a grooved wheel with a handle was mounted near it. A gut line round this wheel turned the work. Later versions had a small pulley that ran freely on one of the centres and was turned by the wheel with the handle. A small projecting rod was attached to the workpiece and this was engaged by a projection on the free-running pulley.

This hand-operated lathe was called by clockmakers and watchmakers a throw, and it was in common use until recent times. Another version had a treadle instead of a handle.

The lathe with a chuck running in a bearing (the headstock) to hold the work was introduced in the eighteenth century, but did not become popular with clockmakers and watchmakers until the twentieth century, because work turned between the dead centres of the turns or a throw was much more accurate than that turned on lathes with bearings that had to have a certain amount of play in them. It was not until relatively recently that this problem was effectively eliminated. Even so, there were diehards still using turns until the last decade or so.

The French introduced the screw-cutting lathe in 1568, but makers of clocks and watches continued to use screw plates for the small parts they needed. The newer tools never provided the accuracy required until they had been improved over many years. The modern screw-cutting lathe with a slide rest was invented in 1794 by an Englishman,

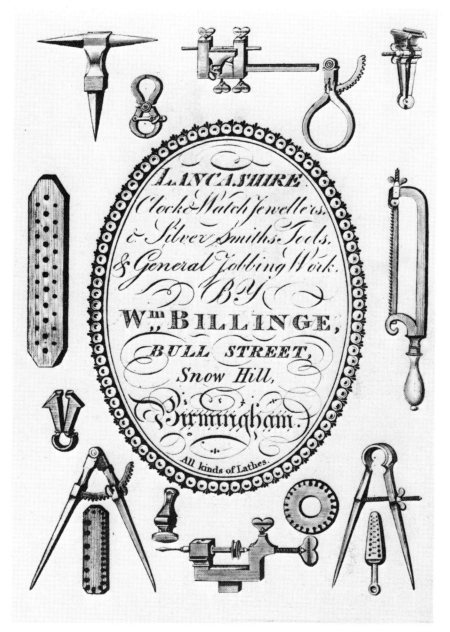

Henry Maudslay, and again it was a long time before it was adopted by the makers of clocks and watches.

For makers of timepieces, the most valuable innovation was the dividing plate, which enabled blank wheels to be marked out accurately before the teeth were filed. It was a circular plate with a series of small holes around the edge corresponding to the common numbers of teeth used on wheels.

To become a dividing engine, the dividing plate was mounted on an axle in a frame. A pin on the end of a lever attached to the frame dropped into a hole to prevent the wheel from rotating. It could then be used to turn the dividing wheel one hole at a time. The blank was attached to the dividing plate and a scriber marked each position where a tooth should be filed.

Above : A page from Wrightson and Webb's Directory of Birmingham, 1835, showing miniature anvil and vice and two sets of callipers at the top. On one side is a drawplate for making wires, and on the other a hacksaw. At the bottom are two pairs of dividers, a throw and a planishing hammer head. The three plates with holes are screw plates for threading rod.

In the mid-seventeenth century, it may have been Dr Robert Hooke, the Royal Society experimenter, who realized that the scriber could be replaced by a cutting tool – a reciprocating or rotating cutter – that would actually cut the teeth, but it took a long time for such an advanced idea to be accepted, and even three-quarters of a century later, John Smith, in his *Horological Dialogues*, implied that hand-cutting was still very common when he commented that no man could cut teeth 'down by hand so true and equal as an engine does'.

In the eighteenth and nineteenth centuries, hand- and treadle-operated wheel cutting engines became universal, and the industries making them flourished in France and especially in Lancashire, England.

Similar engines were used for escape

Above : If a hole has to be drilled in an eccentric position, a mandrel lathe like this is used to position the part being drilled by the three clamps. It was made in the nineteenth century, with adjustable slide tool rest and is driven by a handle through helical gearing.

Left : A wheel cutting engine inscribed 'Knight's Horologium machina 1783'. The indexing wheel at the bottom determines the number of teeth. The pulley drive provides high speeds for the cutting wheel.

wheels, crown wheels, and for pinions, the small driven gear wheels, but for them the Lancashire industry developed a more convenient alternative known as pinion wire, like a very long pinion that could be cut off to the length required for both pinion and staff (axle), for which the teeth were turned off. Pinion wire was produced by simply drawing round wire through a pinion-shaped die. This had for centuries been a common practice of goldsmiths and silversmiths making ornamental wires and was easily adopted by the clockmakers.

The situation in 1747 was described by a contemporary writer in *The London Tradesman* in these words:

The Movement-Maker forges his Wheels of brass to the just Dimensions, sends them to the Cutter, and had them cut at a trifling Expense: He had nothing to do when he takes them from the Cutter but has to finish them and turn the Corners of the Teeth. The Pinions made of Steel are drawn at the Mill, so that the Watch-Maker has only to file down the Pivots, and fix them to their proper Wheels.

The Springs are made by a Tradesman who does nothing else, and the Chains by

Left : The staff of
Ashbourne, Derby,
England, packing
clocks into a crate in
about 1865. The firm
made longcase and
skeleton clocks on a
batch programme
which, as in other small
factories of the time,
had some unrecognized
elements of mass
production.

Below : A mezzotint of
John Ellicott, F.R.S.,
by Robert Duncarton,
after a painting by
Nathaniel Dance.

another: These last are frequently made by
Women, in the Country about *London*, and
sold to the Watch-Maker by the Dozen for
a very small Price. It requires no great Ingenuity to learn to make Watch-Chains,
the Instruments made for that Use renders
the Work quite easy, which to the eye
would appear very difficult.

There are Workmen who make nothing
else but the Caps and Studs for Watches,
and Silver-Smiths who only make Cases,
and Workmen who only cut the Dial-
Plates, or enamel them, which is of late
become much the Fashion.

When the Watch-Maker has got home
all the Movements of the Watch, and the
other different Parts of which it consists,
he gives the whole to a Finisher, who puts
the whole Machine together, having first
had the Brass-Wheels gilded by the Gilder,
and adjusts it to proper Time. The Watch-
Maker puts his Name upon the Plate, and
is esteemed as the Maker, though he has
not made in his Shop the smallest Wheel
belonging to it.

The same was true of clocks, although it
was more common for specialist firms to make
complete movements and supply them to
'clockmakers' in various parts of the country
with the name of the 'clockmaker' engraved on
the plates and engraved or painted on the dial.
There were several such suppliers, embryo
factories, in London and Birmingham, and
also in various European towns, particularly
in France, where such early factories situated
in Paris, and in the small town of Saint Nicolas-
d'Aliermont, east of Dieppe, made carriage
clocks.

In England, the most active trade clockmakers behind a well-known name were Aynsworth and John Thwaites, who later, with George Jeremiah Reed, became Thwaites and Reed, but there were others, including another London firm, Handley and Moore, as well as Whitehurst of Derby, Walker and Finnemore of Birmingham, Ainsworth of Warrington. In Birmingham also, for a short time in 1772–3, Matthew Boulton, known as the father of the Birmingham metal-working trades, had a clockmaking factory supplying the trade.

Famous makers took advantage of these services, mainly for certain of their bread-and-butter lines, which, incidentally, included repeaters. Among them were John Ellicott, B. L. Vulliamy, Paul Philip Barraud, William Dutton, Francis Perrigal, William Frodsham, Le Coq, George Prior and John Dwerrihouse, all well-known makers of the eighteenth century. Thwaites and Reed still exist, but as part of F.W. Elliott (Holdings), and look after Big Ben, the makers of which, E. Dent and Co., are still in business.

Subdivision of labour and the development of accurate tools to speed production had led to the vital step, the making of interchangeable parts. One man concentrating on one

Above : William James Frodsham (1778–1850), another Fellow of the Royal Society. His firm persisted in watch and clockmaking until the mid-twentieth century and had a small part in its revival after 1945.

Far right : Eli Terry, the American clockmaker who was the true pioneer of mass production methods, at about the same time as Eli Whitney. Terry made wooden clock movements ; Whitney made muskets.

Right : Benjamin Lewis Vulliamy (1780–1854), from an unsigned colour portrait. The Vulliamy family held the Royal warrant through three generations.

particular part which can fit any clock, can make such parts faster, and usually better, than a man who has to make all of the clock. Stocks can be held to assemble or complete clocks when required, saving capital and shortening delivery time.

It was found economical to make clocks in batches of a single type, striking bracket clocks that would repeat on the hour, for example. The plates would be cast and hammered flat, drilled together to form the pivot holes for standardized, interchangeable wheels and pinions, escapement, various levers, and so on. When the parts were available, the clocks would be assembled in batches of any number, which also saved time. The numbers of clocks produced by old workshops indicate that they must have used such methods. Thomas Tompion, for example, must have done so.

THE INDUSTRIAL REVOLUTION

The power used was still the muscle of man. The next major stride came with the Industrial Revolution, sparked off by James Watt's invention that made Newcomen's primitive steam pump into an efficient source of extra power, and turned Britain from a rural society into the world's first industrial nation. The Industrial Revolution spread to other European countries, and to America, where it had the most explosive effect of all.

A technology was now required to transform the old craft methods into new ones which could be used in a steam factory. Arkwright partly achieved this in his cotton mill in England; he also had the concept of raw materials flowing in at one end and finished goods flowing out of the other, but the first flow production was in a converted flour

mill in England, where, in 1790, a workforce began making pulley blocks for the Royal Navy. The system was invented by Marc Isambard Brunel and Henry Maudslay.

Eight years later, Eli Whitney, in America, obtained an order from the United States Government for 10,000 muskets, to be delivered in 15 months. There was no way in which the delivery time could be met by craftsmen making muskets in the traditional way, one at a time. Whitney designed a musket consisting of a number of quite separate parts, that could be made independently to a fixed standard of accuracy. He achieved

MESSRS. JOHN MOORE AND SONS' CLOCK FACTORY.

Below : John Moore and Sons of London, made orthodox and musical domestic clocks, and particularly turret clocks, during the early nineteenth century in their two-storey steam factory.

this by devising stencils so that holes were drilled in identical positions each time, by stops to limit the sizes of work carried out in lathes, and by filing jigs, so that parts could not be filed under or over size.

To demonstrate the interchangeability, he took a number of bags, each containing ten identical parts, to the Treasury Office in Philadelphia. Asking officials to select any one part from any bag, he assembled a musket before their eyes. The demonstration would be considered childish today, but then it was an industrial miracle, especially as the parts had been made so quickly and by semi-skilled labour.

To follow the trail that led a clockmaker to follow Whitney's lead, it is necessary to go back to 1773, when an Englishman named Thomas Harland went to the American colonies, where he landed in Boston at the time of the Boston Tea Party. He escaped from the tense atmosphere of the port by going south and setting up a workshop in Norwich, Connecticut, where he advertised that he made 'in the neatest manner and on the most approved principles, horizontal, repeating and plain watches in gold, silver, metal or covered cases. Spring, musical and plain clocks; church clocks; regulators, etc.' He

announced that he made watch wheels and fusees for the trade 'cut and finished upon the shortest notice, neat as in London, and at the same price'. He was also a skilled goldsmith, silversmith and engraver.

Harland is believed to have been the first man to have made clocks in America in batches from interchangeable parts. One of his apprentices later trained Eli Terry, but before looking at Terry's remarkable career, another pioneering family that influenced him should be mentioned, the Cheneys. Although wooden clocks were produced by a number of makers, Benjamin Cheney, in Connecticut, was probably the first to make in quantity cheap wooden clocks that kept time.

Cheney's clocks had brass dials, which would suggest that the movement was brass as well, but the movement was of wood and so thick that the hood of the longcase clock had to be extra deep. The movements were crude; they were finished one at a time, the wheel teeth being sawn by hand. The escape wheel was of thin steel.

Cheney, his four sons, and several apprentices, proved that there was a big market awaiting the maker of cheap clocks which would keep good time. His lead was followed

Below : The clock case steam workshop of J. Smith and Son of Clerkenwell Green in 1851. Power was used only to supplement men's muscles. The full impact of the industrial revolution was still to be realized.

by a number of other makers. He gave away nothing of his working methods, however. One of his apprentices, John Fitch, who became one of the early steamboat inventors, wrote that, although Cheney was a 'man of considerable genius' with a head which was 'near double the size of common proportions', he left Fitch almost totally ignorant of clockwork.

Eli Terry (1772–1853) was born at East Windsor, on the Connecticut river, eldest son of a farmer and tanner. He was assured of an education because the Windsor selectmen had, from as early as 1650, decreed that every child and apprentice in the town should be taught to read, write, and follow some useful trade or calling. Terry, at the age of 14, must have shown some mechanical aptitude, because he was apprenticed for seven years to Daniel Burnap, who worked in his own district of East Windsor, and had been the most successful of Harland's apprentices. From Burnap he learned all the skills that enabled him to make the expensive longcase clocks with brass movements of the time, and the much cheaper wooden ones.

Particularly interesting is that Terry also did some training with a 'Mr. Cheeney', who might have been either Benjamin Cheney or his brother, Timothy. The Cheneys were at this time already enjoying success with their cheap wooden clocks.

After his service as a journeyman, Terry set up on his own in Northbury, away from his master, Burnap, perhaps to avoid competition with him, but conveniently, because he had become attached to a girl there, whom he married.

He made longcase clocks with both brass and wooden movements and in 1797 took out his first patent, for an equation clock. It had two concentric minute hands, each of different colour and pattern. One showed mean time of day, and the other, together with the striking, indicated apparent time or sundial time. The patent was the first clock patent issued by the United States Patent Office; earlier ones had been under state legislature.

Top left: The movement of an almost entirely wooden clock by Eli Terry. Only the bell and hammer, crutch, and some levers and pins are made of metal.

Above: An Eli Terry box model, with a largely wooden movement seen through the dial.

His brass movements sold for £10–£15, and the wooden ones for £4, or nearly twice that with an engraved brass dial. He made them one at a time, using a hand-operated dividing engine for cutting wheel teeth, and a treadle lathe or throw.

Life was far from easy for Terry. His son recorded that his resources were so limited and the demand for his clocks so small that, after finishing three or four, he was obliged to set out on horseback to sell them. 'His usual way was to put one forward of the saddle on which he rode, one behind, and one on each side of his portmanteau.'

It is not known how or where Terry got his early ideas for mechanization. It may have been in isolation, but this is unlikely, since men creative in different fields were not then so isolated from each other as they tend to be

Below : An American contribution to clock design, an elegant shelf clock, made about 1816 by Eli Terry, and designed for mass production.

today. One of Terry's acquaintances wrote of wire gauges Terry made that dropped past a piece being turned in the lathe when the desired size was reached, and of wheel tooth cutting by machinery, 'it was said was hinted to him by Eli Whitney'.

In 1802 Terry took what seemed to many a rash step in building a small workshop over a stream running down a hill between Plymouth and what is now called Thomaston, and setting up a water-wheel there to drive machinery. Four years later he was able to turn out clocks in batches of 25.

About four years after that, he was making 200 clocks a year by batch production, more than any other clockmaker in the country. A rival clockmaker Gideon Roberts, in Bristol, Connecticut, gloomily told one of his workmen that the country would soon be filled with clocks and his business would be good for nothing in two or three years. He was right; he was put out of business by machine-made clocks.

This was just about the time when Napoleon's blockade of England cut off many of America's supplies, which encouraged new industries there. Roberts's despair was not shared by Edward and Levi Porter, two brothers living in Waterbury. The first was the former Congregational minister in the town; the second a manufacturer of buttons from rolled brass.

For some reason, these two decided that they could sell 4000 movements for longcase clocks provided that they could buy them cheaply enough. The number was probably more than all the clockmakers of Connecticut had made in the previous five or even ten years.

There was only one clockmaker who could possibly cope with such an order. They went to Terry and offered him the wood – oak and laurel – and $4 a movement, when he was currently charging $24 for his own movements. The clocks had to be delivered within three years.

Terry accepted. His mill was too small to do the work, so he sold the mill to Heman Clark, once apprenticed to him, and bought a larger one, at a place called Ireland, from a carpenter named Calvin Hoadley. For a year, he planned the layout, designed and tested machine tools, and, presumably, at some stage, trained some workmen in the new methods.

It was the best way to approach such a radical concept of production, but it must have required great courage and faith, because he was not wealthy and his wife had just given birth to their fifth child.

clock, rather than a wall or a standing clock, was the answer.

From 1813 to 1814, he fitted out a new mill on the Naugatuck river. The shelf clock he designed was in a rectangular box case 50 centimetres (20 inches) high and only 10 centimetres (4 inches) deep. The wooden movement ran for 30 hours. Some of the first models had the numerals painted on a glass in front of the exposed movement; later ones had painted wooden dials. In the first three years, he sold 4000; five years later, his production had risen to 6000 a year. Eli Terry was again successful in a risky pioneering venture.

Public acclaim of the shelf clock was the death knell of the longcase clock, which by 1820 to 1825 had gone out of production. There was also a shift of centre to Bristol,

Left: One of the popular calendar models by the Ansonia Brass and Copper Company, made under a patent of Eli Terry's in Connecticut, the cradle of mass clock production.

Below: A small shelf clock by an unknown Connecticut maker, between 1640 and 1865. It is 26cm (14 ins) high and veneered in mahogany.

The mill was fitted out with shafts driven by the water-wheel. From the shafts, belts drove saws, drills and lathes. All components were standardized in size and various jigs and stops employed to keep to the standards. Circular wheel blanks from the lathe were set up in dozens for the teeth to be cut at the same time.

In 1808 Terry put through a pilot batch of 500 clock movements. By the end of the year he had finished over 1000 and had taken on an assistant, named Seth Thomas, to help with final assembly and adjustment. Then he took on Silas Hoadley, the son of the carpenter who had sold him the mill. He completed the contract for the Porter brothers in 1809, and sold the business to his two assistants, Seth Thomas and Silas Hoadley, for $6000 the following year.

Terry did not immediately start manufacture again. As before, he took time off to think, and in the meantime bought a house in Waterbury from the Porters (or it may have been made over to him in part payment for the clocks). He was also suffering from gout. But his main preoccupation was with what would follow the longcase clock. He must have known the Massachusetts shelf clock in its various forms, and decided that a shelf

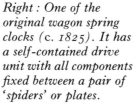

Left : A wagon spring clock, invented by Joseph Ives, and made by Birge and Fuller. A leaf spring, like that used in a cart, drives the clock through a lever and chain driving a fusee on each side.

Right : One of the original wagon spring clocks (c. 1825). It has a self-contained drive unit with all components fixed between a pair of 'spiders' or plates.

where many new clockmaking concerns were established, so that, by about 1840, it had begun to take the lead even over its main European rival, the German industry in the Black Forest, and became the world's most prolific clock-producing centre.

It was a time of intense activity in Bristol, and indeed in other clockmaking centres in Connecticut, including Waterbury and Winsted. Terry was not the only one to experiment with mechanical production, but on his success many more followed suit, among them, of course, his former partners, Hoadley and Thomas. These two stayed in business together for three years, until Thomas sold out to Hoadley and set up on his own.

Terry's patent protected him to some extent, but not from such makers as Silas Hoadley, Mark Leavenworth, Joseph Ives and Chauncey Boardman, who managed to design clocks that were similar, but not near enough to Terry's to warrant legal action.

Seth Thomas moved in the other direction. He obtained a licence in 1818 to make Terry's shelf clocks on payment of a fee of 50 cents a clock. Terry went on to improve his clock several times, and Thomas was licensed for certain improved models. After obtaining a patent for the improvements in 1822, how-

ever, Seth Thomas turned down Terry's offer of a renewal of the licence but went on to copy the clock just the same, which led to a court case between the two men.

For some years before he died, a rich man, Terry left the actual manufacturing of clocks to his sons, and concentrated on invention and improvement. He experimented with the mass production of brass clocks, as did others, but the cheapness of the wooden clock made it too serious a competitor. Heman Clark, Terry's one-time apprentice, did set up a business making brass clocks, as did Joseph Ives, and one of Terry's sons, Silas B. Terry, but not on the scale on which wooden clocks were manufactured. (Clark made a version of a brass clock designed by Eli Terry.)

Nevertheless, Joseph Ives had a factory in Bristol in 1819 devoted only to brass clocks, of which 2000 a year were made. The report containing this information also recorded that half the town's inhabitants were engaged in mechanical manufacturing, mainly of wooden and brass clocks, and of tinwares.

Joseph Ives (1782–1862) was enthusiastic about lantern pinions, which he made with rollers, that acted as roller bearings and reduced friction. Like other makers, he tried to conserve brass, and his frames were a series

Right : Brass was difficult to obtain for early American brass clock movements, so the metal was reduced to a minimum, as in this example.

of straps or a pierced sheet.

Present-day collectors associate the name of Joseph Ives with the wagon spring clock. Ives was not satisfied with the ordinary coiled clock spring. Spring-driven shelf clocks were in competition with the more accurate weight-driven shelf clocks. He experimented with the reintroduction of the fusee, in an unusual shelf clock of hourglass shape, but did not use a coiled spring. Instead there was a U-shaped flat strip of tempered steel that provided the power for both going and striking trains of gears, both of which had fusees.

The horseshoe driving spring was devel-oped from his earlier wagon spring. In this the coiled springs were replaced by a series of flat bars of diminishing length, bolted to-gether in the middle, like the elliptical spring used in a wagon suspension. When the two trains of the clock were wound, each end of the spring was pulled upwards. The idea was patented in 1845 and shortly afterwards mod-ified by intermediate barrels between the spring ends and driving barrels. Ives called the invention his 'patent elliptical spring', and on the label of Birge and Fuller, who made it under licence, it is called 'J. Ives' patent accelerating spring'.

The trend started by Eli Whitney, the mass-production pioneer, of taking on new and inexperienced workmen, because he found them easy to train 'rather than to attempt to combat the prejudices of those, who had learned the business under another system' had its full effect on Bristol by the 1840s. There were hardly any apprentices, and only a handful of mechanics who knew how to make a complete clock.

An interesting aspect of the times was that little business was done by cash. Earlier, when clockmakers peddled their own clocks, they often had to settle a deal by barter; there are old account books testifying to this. For example, one of Daniel Burnap's memoranda states: 'Mr. Austin Phelps Simsbury to have a Clock the Case 7 feet & 7 inches high he agreed to make me a Hors Waggon compleet & pay 50s in Cash for the Clock.'

These methods continued. Firms in Bristol would balance their accounts with each other at certain intervals and settle up with goods, perhaps labour as well, and cash. Even workmen in the factory would draw goods and food from local stores on account, which was settled against their wages.

There were many specialist manufacturers in Bristol, and to be a manufacturer anywhere else only needed an assembly bench. This is how Chauncey Jerome (1793–1860) started, before he was responsible for the final blow to the wooden clock. He bought movements and cases from Connecticut and performed the simple job of putting them together in Richmond, Virginia. Jerome once worked for Eli Terry and claimed to have made the first pillar and scroll clock case for him. He was a joiner and salesman and was in business with his younger brother, Noble, who was a clockmaker.

The brothers went to Bristol in 1821. They began making shelf clocks and were soon among the bigger manufacturers. In 1839 Noble Jerome obtained a patent for a cheap brass clock movement running for about 30 hours. One way in which money was saved was to revert to the early count wheel in place of a rack to control the striking. To avoid the nuisance of having to make the clock strike through the hours, if striking and hands became out of step, the count wheel was friction-mounted so that it could be turned.

Chauncey Jerome claimed the credit for the invention, when he wrote 'What I originated that night [in 1838] on my bed in Richmond has given work to thousands of men yearly for more than twenty years, built up the largest manufactures in New England,

and put more than a million dollars into the hands of the brass makers.' Jerome also remarked that wooden clocks were slow to make properly, that wood for wheels and plates was difficult to get and that it took a year to season the wood.

The Jeromes's activities had repercussions that shattered the relative peace of the centuries-old clock, and then watch, industries of Europe.

The huge American production of cheap brass clocks was the start of world domination by the American clock industry, leading eventually to the decline of the British industry and a serious crisis in the German. American mass-produced watches had a similar profound effect on the Swiss industry. After the Swiss had adopted American ideas, they in turn gradually drove the British and American industry into the ground. British craftsmen and management were not able to adapt themselves to mass production, although they had pioneered the Industrial Revolution.

The Jeromes were so successful in making and selling brass clocks in America that they managed to reduce the unit cost to the remarkably low figure of only 70 cents for a movement. Other manufacturers became

Above: Chauncey Jerome, the American producer of cheap American clocks with brass movements, swamped the English market with them at unbeatable prices.

*Right : One of
Chauncey Jerome's
eight-day wall clocks,
now a collector's item
in several countries, but
for a long time sought
after only by Americans.*

interested in their methods and copied them.
Even the conservative Seth Thomas had his
nephew trained in their factory.

This perhaps was what led Chauncey
Jerome to look for new fields to conquer. In
1842 he made up a consignment of clocks for
England, but was laughed at for his trouble
by neighbouring clockmakers, who, he wrote,
'ridiculed the idea of selling clocks to England
where labour was so cheap'.

Chauncey Jerome sent his son, also named
Chauncey, and a young man named Epaphro-
ditus Peck, to England to sell the clocks at
$20 each. They were invoiced at only $1.50
each, however, which an import levy of
20 per cent brought up to $1.80.

Customs and Excise officers at once thought
the Jeromes were dumping the clocks at
under-manufacturing cost to soften up the
market before a real assault on it. $1.50 was
much too low a wholesale price for a $20
clock. The Customs took action under the
law of the time by seizing the clocks and paying the invoiced price plus ten per cent. They
thought that the Jeromes would make a
heavy loss, which would discourage them,
because they were certain that clocks could
not possibly be made at these prices. But
Jerome recorded, 'They paid cash for this

*Left : An early and
rather crude circular
issued by Chauncey
Jerome for the English
market. He operated
from a warehouse in
Liverpool.*

*Right : A steeple type
of shelf clock, which
became very popular in
the USA after about
1845. It is the
transatlantic form of
Gothic style.*

Left : One of the watchmaking shops at the First Soviet Watch Factory in Moscow. It was started with the purchase of a complete American watch factory, the Dueber-Hampden Watch Company, in 1930. The industry is now very large and exports to a number of countries including Britain and the USA.

cargo, which made a good sale for us. A few days after, another invoice arrived which our folks entered at the same prices as before; but they were again taken by the officers paying in cash and ten per cent in additions, which was very satisfactory to us.'

Customs must have wondered what was happening to them. They were buying clocks at a rate that was going to produce big problems for them, so they let the next consignment through. That opening was followed up not only by the Jeromes but by a number of other manufacturers. Exports expanded so rapidly that by the 1860s the Americans were exporting to over 30 countries.

Despite a serious fire in one of their factories in 1845, by 1850 the Jerome concerns had become the biggest clock manufacturers in America. Then Jerome went into a financial partnership with Theodore Terry, who had also suffered from a bad fire, and the great showman T. P. Barnum. The partnership proved more disastrous than the fires. In 1855 they were bankrupt. The great Jerome empire was no more. According to one account, Barnum boasted that he could put everything right, which enabled him to get most of the assets into his own hands. He then disposed of them and claimed that he had been ruined by the Jerome company. Chauncey Jerome, after attempts to get started again, had, in his sixties, to go back to work for wages in the traditional white apron at the bench. He died at the age of 67.

From these rumbustious beginnings sprang the huge American manufacturing concerns of the second half of the nineteenth century, the Ansonia Clock Company, the New Haven Clock Company, the Seth Thomas Clock Company, the L. Gilbert Clock Company, the E. Ingraham Clock Company, the Waterbury Clock Company, and the E. N. Welch Manufacturing Company, which became the Sessions Clock Company. They produced a wide selection of clocks, far beyond what had become the traditional American designs, and including copies of French ormolu clocks (but made in wood or cast iron) and of styles from the Black Forest, alarms of many types, all kinds of novelty clocks, 'Viennese' regulators, and many others.

The Ansonia Company became particularly well known for imitation ormolu and novelty clocks, which were sold all over the world. Before the First World War, competition became too fierce, and, while other companies tried to maintain realistic prices, the Ansonia Company cut them to the extent that they made heavy losses and had to go out of busi-

ness. In 1929 the bulk of the factory machinery and certain tools and dies were sold to the Russian Government and became the foundation of the clock industry there. The Russians also bought the Dueber-Hampden Watch Company in 1930. The plant and tools were shipped from the United States to the Soviet Union, accompanied by 21 former employees of the company, who helped to set up the First Soviet Watch Factory in Moscow.

The world depression after the First World War, and then the Second World War, put all the giant clock companies out of business,

Above : One outward expression of mass production in clocks, particularly in France, was to use industrial and engineering symbols in clock cases, to which standardized drum movements were fitted. Another expression was in mystery clocks, where the means of operation was disguised.

Left : The French carriage clock, derived from the pendule d'officier *used by Napoleon's officers, was one of the earliest to be produced in quantity in Paris, Besançon, and a district near Dieppe. Some were still individually made, however, like this repeater by Barwise, London.*

with the exception of the Seth Thomas Company, which eventually became a division of General Time and then, in 1970, of Talley Industries of Seattle, Washington, which also controls Westclox.

The Black Forest clock industry, which depended heavily on exporting, soon felt the impact of American competition, especially as the Americans were copying German designs and selling the clocks much more cheaply in the same markets. But Black Forest clockmakers were apparently not quite so steeped in the traditions of apprenticeship and traditional methods as the older clockmaking centres in Europe. They were gradually persuaded, mainly by the example of one man, Erhard Junghans, to adopt the deskilled factory system developed by the Americans. The Germans specialized in

cheap clocks, which was probably the main reason why they accepted change more readily than the British, who concentrated on higher quality.

Erhard Junghans (1823–70) was involved with a business making plaited straw for hats, and then with a mill that was converted to pressing brass surrounds for picture-frame clocks, when the American Civil War caused the price of straw to double. At about the same time, Erhard's elder brother, Xavier, a cabinet maker who had emigrated to America, was persuaded to order some of the new clockmaking machinery and come home to join the family in a business making complete clocks.

Apparently Erhard Junghans intended from the beginning to employ the American system, but he had to convince those who worked for him and supplied him that success depended upon interchangeability. First production was of a marine type of clock with a balance, a mantel clock with pendulum, and a striking wall clock with pendulum, all being spring-driven.

The clocks were rather solid, and methods of production still relatively primitive, when the factory went into operation in 1864. It was nearly two years before the firm managed to change over to American wheel cutting methods, and it was not until 1870, the year Junghans died, that it was really Americanized in style of production and type of clock movement.

In France the Industrial Revolution seems to have had more influence on the design of clock cases and movements than on production methods. During the last thirty years of the nineteenth century, clocks representing

Left : Clock factories, like this one belonging to S. Smith and Sons at Cricklewood, near London, were turned over to fuse making in the First World War, as were such factories in other combatant countries. Smiths became the biggest makers of clocks and jewelled watches in the UK, but no longer make watches.

Left : In the age of mass production, the case sometimes becomes more important as a means of expressing individuality. This nineteenth-century boudoire clock has a porcelain case and painted figures.

Right : Another example of a standard drum movement, a French striking clock with Breguet hands in a porcelain case, again with children as the motif.

163

working steam hammers, boats, beam engines, and other symbols of industry, became popular. The steam hammer was moved up and down or the beam rocked by the pendulum. Some clocks were like magic lanterns and projected an image of the dial on the wall at night. Early in the twentieth century there were even clocks in the shape of submarines and motor cars with moving parts.

The British industry slowly declined after the first impact from the other side of the Atlantic. Bracket or mantel clocks of good quality continued to be made, with fine fusee spring movements and standard cabinet cases. American imports were disregarded as being trash; indeed the same attitude was perpetuated by collectors and still exists to a large extent in Europe today.

One of the principal manufacturers was J. Smith and Sons, who had a steam factory in Clerkenwell, London. The firm still exists, but now sells non-ferrous metals. They concentrated on bracket clocks, but made some longcase clocks and regulators. The hey-day of the longcase clock was over. A new style of English clock, which became typical of the Victorian era, was about to be launched, the brass skeleton clock. In no way did it stop the American invasion, which was opening up a new market, but it did boost the domestic industry.

The skeleton clock has its plates cut away to show the gear train and escapement. The French probably thought of the idea first, as a number of the master clockmakers, including Le Roy, Lepaute, Berthoud and Lépine, made them in the second half of the eighteenth century. As the movement is intended

Below : When the Americans invaded the English market with cheap clocks, some English makers strove to keep in business by turning to novelties. The most successful was the skeleton clock, which appealed to Victorian taste, especially the monumental type, like the Scott Memorial clock at the bottom left and the York Minster clock below.

Right : Mass production methods were applied to dials, cases, and other clock parts by the Americans. An employee of the Seth Thomas factory here inspects printed clock dials.

Right : Skeleton clocks achieved several stages of sophistication for the top end of the market, while allowing the Americans to flood the rest. This version with a calendar, runs for 400 days at a single winding. It was usual to keep a winding book with signatures and dates of winding for such clocks.

to be seen, there is no case and the clock is enclosed by a glass dome on a plinth.

The English did not copy the relatively simple frames of most French clocks, but introduced much more piercing and fretting of the polished brass plates. Once seen, the English style of skeleton clock is readily recognizable again. Most of them are solidly engineered and have the traditional fusee drive, the spring requiring weekly winding, but some went for longer periods at a winding, a month, or even, in a few cases, a year.

Gothic cathedrals were a favourite theme for the frame, and a number of these and other buildings are recognizable, such as Westminster Abbey, York Minster, Lichfield Cathedral, Milan Cathedral, the Royal Pavilion at Brighton and the Scott Memorial, Edinburgh. Others, of more scrolled form, were in the shapes of hearts, lyres, wheatsheaves, and so on.

Tens of thousands of skeleton clocks were made from about 1820 to within a decade or so of the end of the century by manufacturers in Clerkenwell, London (where J. Smith and Sons made a big variety), Birmingham, Liverpool, Prescot in Lancashire, and Derby. The skeleton clock was almost finished when the First World War broke out.

VIII
THE TECHNOLOGICAL AGE

Most new ideas are unoriginal, unworkable, uneconomic or unscientific. Very occasionally, one is practicable and economic, although it may not even be new. On the very, very rare occasion, an idea emerges that is really unique and brilliant, such was the use of electricity.

The extreme simplicity of the objective in horology is its attraction to the scientist as well as to the engineer and the mechanic. At first, the object was to make a pointer that always pointed to the sun; then, when the sun was found to be somewhat inaccurate, to make a hand go round in equal hours. The ultimate aim was to make a better timekeeper than the stars.

ELECTRICITY IN CLOCKS

A monumental row between a scientist and a mechanic blew up over the invention of the electric clock. In 1837 a young Scotsman named Alexander Bain (*c.* 1811–77) arrived in London from Edinburgh. He had served his apprenticeship as a clockmaker and now sought work as a journeyman, that is, a working clockmaker who would serve a master clockmaker until he could set up on his own. Electricity was a new subject, but its applications were beginning to be realized and Bain already had an idea for an electric telegraph. The great man in London on the subject was Charles Wheatstone (1802–75), later Sir Charles, who had just, with William Cooke, patented an electric telegraph.

Bain obtained an interview with Wheatstone in order to show him some models and to seek his patronage for developing various ideas. While they were discussing the telegraph, the subject of electric clocks came up. Bain said that he had been looking at the 'beautiful electro-magnetic apparatus in motion at the Adelaide Gallery', and it had occurred to him that the same power might be used, with advantage, in working clocks. 'Very shortly after the idea was conceived, I began reducing it to practice, and have ever since, although much engaged with the twin invention, the Electric Telegraph, been testing and improving it.'

The idea of using electricity for timetelling had already occurred to a number of people. For example, there is a description of a galvanic clock by Francis Reynolds of Hammersmith in the *Philosophical Magazine* of 1814. The problem was to make a practical clock. Bain's idea was for what is now called a master and slave clock system, in which one clock operated a number of dials in different places.

Far left : The detent or chronometer escapement is a classic escapement and is used in this silver-cased clock (1974) by the marine chronometer makers, Thomas Mercer Ltd of St Albans.

Left : Alexander Bain, the Scottish pioneer of electrical timekeeping, photographed in 1874.

Left : Sir Charles Wheatstone with a group of fellow scientists – Faraday, Huxley, Brewster, and Tyndall – in 1876. Despite his eminence, he astonishingly claimed some of Bain's ideas.

Below, left and right : The electric master clock and slave dial made by Wheatstone's company, the British Telegraph Manufacturing Co., in about 1870. A pendulum with a coil for a bob behaves like a generator of alternating current to drive the slave.

With another clockmaker, named Barwise, he obtained a patent for the system on 10 October 1840. Immediately afterwards, however, in November, Wheatstone read a paper describing the same kind of system to the Royal Society and exhibited an electric clock.

As soon as Bain heard about the lecture, he was convinced that Wheatstone had stolen his idea. In May 1842, Wheatstone's clock was exhibited in the library of the Royal Institution in Albemarle Street, London, and reported upon in the *Literary Gazette*, which so infuriated Bain that he wrote a letter to the editor of the publication. Alas, Bain was not very literate, and the editor was unkind and supercilious. He published it with all the misspellings and ungrammatical construction and added a footnote: 'Printed *verbatim et literatim. Fiat justitia, ruat coelum.*'

In the next issue, Wheatstone, who had been professor of physics at King's College, London, since 1834, slaughtered Bain in print, referring to him, untruthfully, as 'this working mechanic formerly in my employ' and issuing the countercharge that Bain had stolen his ideas. He asserted that he, Wheatstone, had discussed electric clocks with various engineers, including Brunel, early in 1840, although he had not applied for patents.

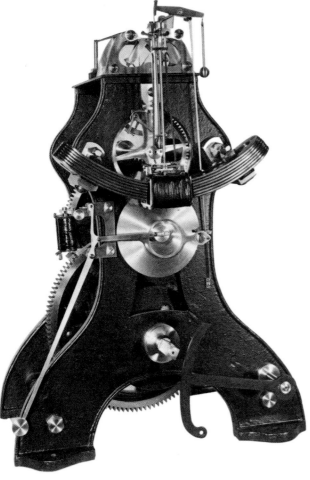

In those days the dice were laden against the mechanics, but Wheatstone took on a determined man when he tackled Bain. Later, Bain and Barwise sued him and won. The *Mechanics' Magazine* was naturally on their side and published a congratulatory message, with the comment, 'we earnestly trust that neither pirates nor professors will again annoy him'. This was in 1844.

In an 1846 issue of the *Mechanics' Magazine*, it was reported that:

The ingenious inventor [Bain] has directed his attention to the contrivance of a plan whereby a system of uniform time may be established thoughout the country . . . Mr. Bain placed the pendulum of one of his Electric Clocks at the Edinburgh station of the Edinburgh and Glasgow railway; and the works of a common timepiece at the Glasgow terminus were set in motion simultaneously with it, by means of the electric wire connecting the two stations. . .

Bain made grandfather clocks in which the pendulum bob was a coil of wire, or two coils, swinging over a bar magnet. It was operated by an earth battery. He explained in a booklet how he discovered the idea and how it was operated.

If we place a sheet of zinc and another of copper in the ground (each sheet having a copper wire previously soldered to it) a little distance from each other, and a few feet deep, so that they are perfectly imbedded in the moist soil, we have, by this simple arrangement, a source of electricity, and if the sheets of metal are about two feet each we shall have amply sufficient to work a clock.

There are still some of these clocks about.

GREENWICH MEAN TIME

By the nineteenth century marine chronometers had achieved a high degree of accuracy. However, there was still need for a time signal of some kind by which to set the ships' chronometers before sailing.

John Pond, Astronomer Royal, solved the problem in 1833 by installing a time ball on top of Greenwich Observatory. It was a wooden ball, about 90 centimetres (3 feet) in diameter and covered with leather, which slid up and down on a mast 4.5 metres (15 feet) high. At 12.58 the ball was wound halfway up the mast as a warning and at precisely 01.00 it was dropped by a trigger release. Nineteen years later, an electrical

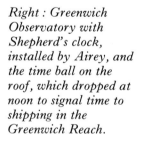

Right : Greenwich Observatory with Shepherd's clock, installed by Airey, and the time ball on the roof, which dropped at noon to signal time to shipping in the Greenwich Reach.

Right : A longcase clock in the Gothic style by Alexander Bain. The pendulum bob coil and the sliding switch on the pendulum rod can be seen clearly.

Below : An electrically-operated time ball which was installed in The Strand, London.

device was installed so that the Observatory mean solar time clock could release it automatically.

The visual signal of a falling ball was much more accurate than the old aural signals of a striking clock or a time gun. Time balls were installed in many ports subsequently, including Deal, Brighton, Devonport, Portsmouth and Portland. There was also one in The Strand, London, for the many chronometer makers who worked there.

The electric telegraph revolutionized time distribution. One of the conditions laid down by the Astronomer Royal for the Big Ben clock was that it had to be linked to Greenwich by electric wire for time checks. Wires were also laid from Greenwich to the Post Office, which became and still is a vital link in the distribution of time. Originally, private companies, such as the Standard Time Company, were involved too. This concern supplied time to individual clockmakers and watchmakers in the manufacturing district of Clerkenwell in London. They would fit a special clock on which the electric signal forcibly corrected the hands. Clerkenwell chronometer makers also had a personal service supplied by a Miss Ruth Belville, who daily journeyed from Greenwich with her

lems of the American railroads, which could not keep Washington time over a network some 4000 kilometres (2500 miles) wide. In 1884, an international conference decided to adopt the meridian through Greenwich as zero, and therefore to adopt Greenwich Mean Time, and to divide the world into 24 time zones each an hour apart. The USA made the proposition and only Ireland and France opposed it.

THE SYNCHRONOME CLOCK
With the spread of industrialization, good timekeeping became increasingly important. Makers were still obsessed by compensation for temperature changes. It had been discovered that chronometers and watches with hairsprings and bimetallic balances which altered in diameter to correct them could

Left: The one o'clock time gun at Edinburgh Castle. The clock tripped a weighted lever which fired the cannon.

Below: A Swiss chaise or travelling clock-watch with an alarm, made by Jean Baptiste Duboule of Geneva (1615–94), who also engraved it. It displays the day and date, age of the moon and its phase, as well as signs of the zodiac and seasons.

Arnold chronometer which had been set to time against the Observatory clock.

There was still another problem. As the sundial was used to set the clock, different towns kept different local times. This was not much of an inconvenience as long as the chief method of transport was the horse-drawn coach, as a coach could carry its own time. Special coaching watches were made for the purpose. There were also special locked watches for the mail guards on stage coaches. Serious problems arose with the spread of the railways, however. Crashes occurred because of confusion over timetables caused by the absence of a national time. The railway companies were forced to employ their own 'railway time', which was based on the local mean time at Greenwich. They issued conversion tables for the main stations in different parts of the country, so that the grandfather clock in the station hall or the large-dialled tavern clock (later called an Act of Parliament clock) could be kept at railway time.

In 1880, mean solar time at Greenwich Observatory (GMT) was established by law as the official time for all parts of Britain. In 1876, time zones had been devised by Sanford Fleming to solve the timekeeping prob-

only be adjusted for two temperatures, and between these was a middle temperature error. To overcome the error, a multitude of auxiliary balances was invented. Then temperature error was solved from an unexpected direction, by a scientist named Charles-Edouard Guillaume.

Dr Guillaume was a Swiss who became director of the International Bureau of Weights and Measures in Paris. He carried out some experiments in about 1895 which resulted in his producing an alloy of 35.6 per cent nickel with steel that did not vary in length to any appreciable extent with changes of temperature. He called the metal invar, from *invariable*, and it soon became the invariable pendulum rod for high-precision clocks. Later, Guillaume invented a hairspring alloy which solved the timekeeping

problem for watches. He called it elinvar, from *élasticité invariable*, because its elasticity was unaffected by temperature changes.

For his achievement, Guillaume was awarded a Nobel Prize in 1920, the only man ever to have received one for a contribution to the accuracy of the timekeeper.

Invar and elinvar eliminated the need for many complicated devices and took the clockmaker another step on his path towards isochronism. A pendulum or a balance that swings in equal times regardless of any disturbance is said to be isochronous. Perfect isochronism is impossible. Every timekeeper has some imperfection caused by impulsing it to keep it going and using it to show the time. There are circular errors in pendulums, positional errors in watches caused by movement of the centre of gravity of the balance

Near right: The continuous distribution of time other than by time signals was of growing importance in the nineteenth century. It was solved in Paris for many years by a system of pneumatic pipes which operated street clocks by pulses of air from a master clock.

and spring, temperature errors and barometric errors. Barometric errors are caused by changes in the air's friction, movement of air caused by the pendulum, and flotation effects on the pendulum bob, with changes in air pressure.

Invar leads us back to the quest for the ideal clock. An invar pendulum and the application of electricity to the gravity escapement led to the ultimate in pendulum clocks.

The man who took the first step this time was Frank Hope-Jones, of whom Sir Charles Boys remarked:

> He writes as the 'high priest' who knows that his religion is the only true religion and his faith the only true faith . . . he preaches the one doctrine as to the way to follow to attain the perfect clock; all other ways are vain, and lead, if not to perdition, to less perfect results. Like St Athanasius, his faith is clear and emphatic, and the conclusion of 'St Hope-Jones' is like that of his predecessor and in almost the same words, 'this is the clock faith; which except a man keep faithfully, he cannot be safe'.

Hope-Jones in fact quoted another clockmaker, H.M. Astronomer Sir David Gill, as expressing his ideal:

> . . . to maintain the motion of a free pendulum in a uniform arc, when the pendulum is kept in uniform pressure and temperature, and to record the number of vibrations which the pendulum performs, is to realize the conditions which constitute a perfect clock.

The detached gravity escapement pendulum clock devised by Hope-Jones was simple, but incorporated a number of fundamental principles. The small impulse to keep the pendulum swinging was delivered when the pendulum was at the bottom of its swing, the zero point, and it was delivered near the top of the rod, not to the bob. It was applied gradually and every half-minute instead of at every swing or every other swing of the seconds pendulum, as with most other escapements. The impulse was provided by a lever falling under the influence of gravity, so that it was always constant. The lever was reset by a special switch, designed by Hope-Jones, which always gave good electrical contact for a positive period.

The pendulum turned a tiny wheel whose only work was to count the swings and trip the gravity lever every half-minute. The clock hands were turned electrically every half-minute by the same electrical mechanism that reset the gravity lever. This type of clock, the Synchronome clock, and others of a similar nature, are called master clocks today. They operate many slave dials in large buildings, time-recorders for signing on and off work, and programmes to operate signals for control of factory processes.

Near right : Most towns kept to their own local time, so the Royal Mail had to carry its own time, in the form of a sealed mail guard's watch. This one was used in Ireland in about 1830.

Far right : An unusual eight-day verge travelling clock by Johannes van Ceulen, The Hague, with an alarm, wound by pulling a cord. The handle swivels.

The greatest feat of the Synchronome clock came with the collaboration of Hope-Jones and a civil engineer, W. H. Shortt. In 1921 Shortt hit on the idea of using two pendulums that were precisely synchronized. One did no work at all. It was given a tiny impulse at intervals to keep it swinging. Its function was to keep as accurate time as possible. The second pendulum did the small amount of work involved in turning the counting wheel, and the electrical gear operated the dial or dials. The first pendulum, known as the master or free pendulum, was adjusted precisely to time, and the other, the slave pendulum, to run slightly slow.

The two were linked by a device called a hit-and-miss synchronizer, which kept the slave in step with the master by speeding it up when necessary. An extremely simple but ingenious gravity arm impulsed the master pendulum in the middle of its swing, despite the fact that the gravity arm may have been released a fraction early by the slave. The same arm dropped at a time determined by the master, and this drop provided the synchronizing signal to keep the slave in step.

The pendulum rods were made of invar, which did not vary in length to any appreciable extent in different temperatures, and the master pendulum was mounted in a vacuum chamber. The Shortt free pendulum clock was set up in most of the world's observatories, and kept time to an accuracy hitherto unknown. Even over long periods of several years, most would not vary more than a fraction of a second a year after the very small rate correction had been applied.

Before Shortt, the standard observatory clock was an ingenious mechanical pendulum clock designed by Dr Sigmund Riefler, of Munich. Below the escape wheel was the anchor of the dead-beat escapement. The suspension spring for the pendulum was fastened to the anchor, below its pivot. Thus, when the pendulum swung, it moved the anchor by means of the suspension spring, and was impulsed in the same way, through the suspension spring. In other words, the suspension spring was also used as a spring *remontoire*. Riefler was the first to put a clock in an airtight case at constant pressure, an idea that was copied by other precision clockmakers, including Shortt. He was able to do so by using an electrical winding device invented by Hope-Jones and Bowell in 1895.

The great achievement of Shortt's clock, with its inherent accuracy, was to measure another variation in the timekeeping of the earth. This is nutation, a kind of nodding action in the earth's gyroscopic rotation, which affects its timekeeping by about 20 milliseconds every 15 days, and causes another variation of about one and a fifth seconds every 18 years.

THE SYNCHRONOUS CLOCK

At the opposite end of the horological scale was the synchronous clock with no inherent accuracy. Henry Ellis Warren invented the synchronous electric motor in 1918 and set up a factory in his home town, Ashland, Massachusetts, to manufacture a clock called the Telechron, which employed it. The synchronous motor was a fundamental invention and basically simple. There is a stator wound with coils to which an alternating current is applied. The rotor is arranged so that it runs in step with the electric frequency.

Far left: A tavern clock by Thomas Moore, Ipswich, (1720–1789). Such clocks were also called Act of Parliament clocks because they were often consulted when Parliament imposed a tax on clocks and watches.

Left: A regulator by B. L. Vulliamy (1780–1845), the Royal clockmaker, which has Graham's dead-beat escapement and his mercury pendulum temperature compensation.

Far right : The first practical free pendulum clock was made by W. H. Shortt, but he had been anticipated by R. J. Rudd, who made one and published details of it in 1899.

Below : Art nouveau and art deco cased clocks have become collectors' items in the 1970s. This Cymric clock for Liberty, was made in the 1920s.

This meant that if the frequency of mains electricity were kept in step with an accurate clock, any synchronous motor operated from it would keep time.

A synchronous motor driving hands and operated from the mains is not really a clock; it is a meter, a slave dial following the mains frequency. Its accuracy depends on how much responsibility the electricity authority accepts for this task of keeping the frequency in step with mean time.

In the United States the initial sales of the synchronous clock were on the eastern seaboard, where the electricity companies divided the responsibility among themselves. In a 24-hour period it was corrected by six controls, one after the other. In the United Kingdom it was controlled by the electricity companies, and is now by the Central Elec-

tricity Board, through the national grid. A control engineer compares time on two dials, one showing GMT and the other frequency time, which can be adjusted to GMT by speeding up or slowing down the turbo-generators producing the alternating current.

The firm of S. Smith and Sons, who had started as retail jewellers in The Strand, London, and became (and still is) Britain's biggest manufacturer of clocks and watches, took up manufacture of the synchronous clock and was followed by others. The mains clock became very popular both as a mantel clock and as an alarm, and later its use spread to large public clocks. Indeed, some valuable historic clocks have been destroyed over the years in the name of electrifying the movement.

The Second World War dimmed the popularity of the mains clock somewhat, because load-shedding and cuts in the electricity supply meant that the clocks were often very slow. Non-self-starting versions frequently stopped.

In time of war, clockmakers, watchmakers and jewellers are immediately put to work making fuses, but at the start of the First World War the British clock and watch industries were in a poor way. After the war, Parliament, having realized to what extent defence depended on these industries, applied what were called the McKenna duties to help build them up again. Then, one night, during a dinner of the Worshipful Company of Clockmakers, news came through that the duties had been suspended by the new, pacifist-inclined Labour Government.

The decision gave an opportunity to National Socialist Germany, under Hitler. German clocks were dumped in Britain at well below manufactured cost, with the aim of destroying the British clock industry and with it the fuse-making capacity of Britain. A similar attempt was made on the French industry. When war began, the anti-aircraft shells fired at Luftwaffe bombers had no fuses in them. Fuse mechanisms had to be flown in with great difficulty from Switzerland. France had managed to keep more of her clockmaking industry intact, although it did the country little good because the Germans quickly broke through the Maginot line.

After the war, once more the British Government resolved not to be caught again in the same way, and provided factory space for clockmaking and a substantial loan to encourage the watchmaking industry. The clock industry was quickly set on its feet again,

turning out 8.5 million clocks in the first full year. This renaissance was led by S. Smith and Sons and backed up by other factories, notably those run by two one-time refugees from Russia, Elias Buerger and Dr Andrew Gershoon Colin, who were friends but also shrewd competitors. American companies were also encouraged to set up in Britain. Westclox did so, and have still a factory of world importance in Scotland.

Smiths concentrated on improved design and production methods, especially of alarm clocks, made a wide variety of other clocks, and continued making synchronous clocks, as did some other firms, such as Metamec and Sangamo-Weston. Sales of mains clocks swelled as the electricity authority gradually overcame the demands of industry for power after the war, which delayed a return to a

Below : Some exposed turret clock hands need considerable effort to move them in windy weather, so I. H. Parsons and A. E. J. Ball devised this waiting train movement for Gents in 1907. The electrically-maintained pendulum drives the clock hands over a minute division in about 27 seconds and is then disconnected until, exactly on the minute, a master clock starts the sequence again.

Below : After the Second World War, the British government attempted to revive both clock and watch industries. One of the key firms involved was S. Smith and Sons, who had been very successful with synchronous electric clocks, and then expanded mechanical clock and watch production. This is part of the Cricklewood factory after the war. They are still big clock producers.

good time service through the mains. But a strong competitor to synchronous clocks was over the horizon.

THE BATTERY CLOCK

The development of a dry battery, invented by Alessandro Volta in 1801, which gave consistent power output for a long time and had a long shelf life, made the battery clock possible. (Strictly, a battery is a linked series of cells. Those used to drive clocks and watches are nearly always single cells and should be called power cells, but the word battery is commonly used.) The first dry batteries could run a clock for about a year; then, as clock and battery designs improved, for up to about five years.

In the 1950s the French, who had been pioneers of electric battery clocks with their

Bulle pendulum and Eureka balance wheel clocks, developed the A.T.D., using the principle of the bar magnet pendulum swinging through a fixed coil. It was similar to the Bulle, having a Hipp toggle, an ingenious device that made the circuit to provide the pendulum with an impulse only when the pendulum began to lose arc. In other words, power was supplied on demand. This small and accurate pendulum clock was developed by the Germans to become a 1000-day battery clock, which needed nothing but a very occasional adjustment over this period. The Germans already made a mechanical clock that ran for 400 days and had been developed, since the first model of about 1880, into a very popular clock. It had a torsion pendulum (twisting instead of swinging) and stood in a glass shade. These early commercial battery

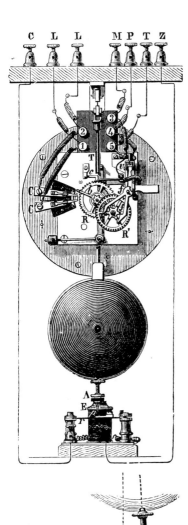

C L L M P T Z

mains synchronous clock. Indeed, it was soon making big inroads into mains clock sales.

Another place where the electric clock gained an early foothold was for the motor car dashboard clock, where, because battery power was always on tap, it was particularly suitable. Smiths developed one that involved the principle of driving the clock through the balance wheel direct. The original idea was beginning to take hold again.

The battery clock had the reliability of a mechanical clock, better timekeeping, and the convenience of requiring little attention. It seemed to mark the limit of development when an invention in another field made another big stride possible. This was the transistor switch.

THE TRANSISTOR

No matter how well made a mechanical switch may be, there is inevitable corrosion and erosion caused by the passage of current and mechanical wear. The transistor was potentially more reliable, very much faster in operation and more accurate. As it completely eliminated arcing, it had no moving parts or contacting surfaces to wear or create resistance. Early versions did not always live up to these promises, but the problems were in design and production and were gradually overcome.

The transistor, invented in 1948, was first applied to the 1000-day clock and simply replaced the mechanical switch – the Hipp toggle. But that introduced another problem: how could the pendulum trigger the transistor switch? This difficulty was solved by providing a second, smaller, coil next to the

Above: Matthäu Hipp of Neuchâtel invented an ingenious toggle switch which much improved the performance of the electric clock because the pendulum was impulsed only when the arc fell off.

Right: An electric clock invented in 1906 and made by the Eureka Clock Company. It is controlled by an electromagnetically impulsed balance wheel instead of a pendulum.

clocks were driven directly by electromagnetism applied to the pendulum or balance, not by applying it to wind a weight or spring. Yet curiously, it was a backward step that led to the great popularity of the battery clock.

After the Second World War, some manufacturers, Keinzle being one of the most successful, began making mechanical clocks with small mainsprings that were rewound every quarter of an hour or so by a battery operating a solenoid. They were *remontoire* clocks (from the French for a keyless watch), driven by a light spring, frequently rewound to provide relatively constant driving force, and therefore good timekeeping. This type of electric timekeeper was called the cordless clock (without electric cable) in America, which emphasized it as a competitor of the

impulsing coil, through which the pendulum swung. As the magnetic pendulum bob moved through this coil, a small current was generated in it, which triggered the transistor to send a larger current through the impulsing coil. Thus the pendulum controlled its own impulsing to keep its rate steady.

Inventors soon realized that a new principle was involved. The escapement, as it had been known, had been replaced by a transistor and coil. Patent Offices were flooded with claims, mainly for transistorized balance wheel movements. The clock had become partly electronic.

These developments did not mean that the mechanical clock was finished. It still had a big place in the market. However, manufacturers of mechanical and battery clocks in Europe and the United States were surprised in the 1960s by an unexpected competitor from Japan – a clock controlled by a tuning fork. The idea was not new. The Post Office in Britain had used a tuning fork clock as a time standard. Moreover, the tuning fork watch had been invented. The Japanese, however, applied to it another innovation, the magnetic escapement. This had been invented by a Briton, C. F. Clifford, in 1948, and it was used also in some German clocks

and British time switches. There is no physical contact between the escape wheel and vibrating tuning fork, on the tines of which are tiny magnets to control the passage of the escape wheel 'teeth' between them. What the Japanese did was to keep the tuning fork in vibration by means of a transistorized, battery-powered oscillating circuit. The clock was made under licence in other countries, including the United States and Britain.

THE QUARTZ CRYSTAL CLOCK

The practical quartz crystal clock was invented in 1929 by a Canadian, W. A. Marrison, working in the United States. It is based on the principle that a piece of rock crystal cut to appropriate shape, usually a slice, bar or ring (called an Essen ring after its inventor, who later invented the atomic clock), will resonate at a fixed frequency in an electronic circuit. The crystal replaced the pendulum.

Thus the quartz crystal behaves like an extremely rapidly vibrating balance wheel or pendulum, but controls an electronic instead of a mechanical clock system. Most were cut to resonate at one kilocycle a second for observatory clocks, 2000 times faster than the

Right : Three quartz clocks which were used by the General Post Office – the dial shows the time by one and the rates of any two can be compared in the windows below it.

Below : N.P.L.III, the third caesium atomic clock at the National Physical Laboratory, which has an accuracy of one tenth of a picosecond, or one second in 10,000,000,000,000. An N.P.L. second is compared with those obtained by certain other countries and the mean unit obtained is the basis for the scale of International Atomic Time.

observatory pendulum clock. The quartz observatory clock is accurate to about one second in 30 years, but each clock has its own idiosyncrasies, largely due to an ageing effect on the quartz when it is subjected to constant vibrations. Crystals are aged before being put into service and are mounted carefully at nodal points where there is no vibration. They are sealed in chambers kept at constant temperature, pressure and humidity. Nevertheless, as a rule one quartz clock will vary very slightly from another, and it was found from experience that time could be derived most accurately by taking the average of a small group of clocks. A committee of quartz clocks took over the job of providing accurate time.

The development of quartz as the heart of the domestic and of the personal timekeeper is indissolubly linked with the history of the electronic watch, which is dealt with in the next chapter. Quartz clocks are as commonplace today as mantel clocks, alarms and dashboard clocks.

THE ATOMIC CLOCK

However, there was an even more accurate timekeeper on the way that looked like being the ultimate. It was the atomic clock.

The new master timekeeper was the atom itself. Atoms also have their natural frequencies, and a joint research programme between the National Physical Laboratory in England and the United States Naval Observatory established over three years that an atom of caesium vibrates at 9,192,631,770 cycles in one second of astronomical time. The first atomic clock to go into service, in 1955, was designed by Louis Essen and J. V. L. Parry at the National Physical Laboratory. The atomic clock, using the caesium atom as its 'pendulum', could keep time to an accuracy of about one second in 3000 years.

In practice, the atom clock is made to control a quartz clock which controls an oscillating circuit that operates an electric motor to turn the clock hands. Both the United States and the United Kingdom broadcast frequencies based on atomic clocks that are available on a world-wide basis and are accurate to one part in 10,000 million. They are used for accurate time measurements, for accurate frequency standards (in broadcasting, for example) and for navigational purposes. The United Kingdom broadcasts are made from the Post Office station at Rugby on behalf of the National Physical Laboratory, and the American ones from Fort Collins, Colorado.

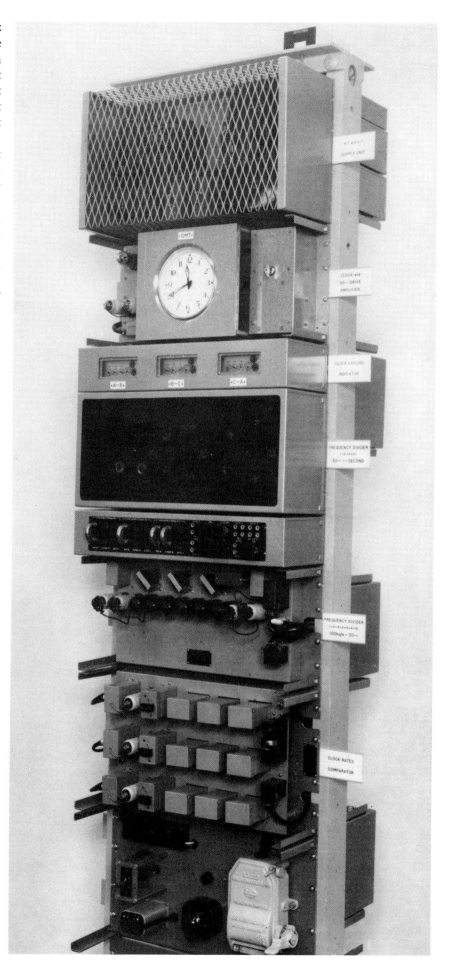

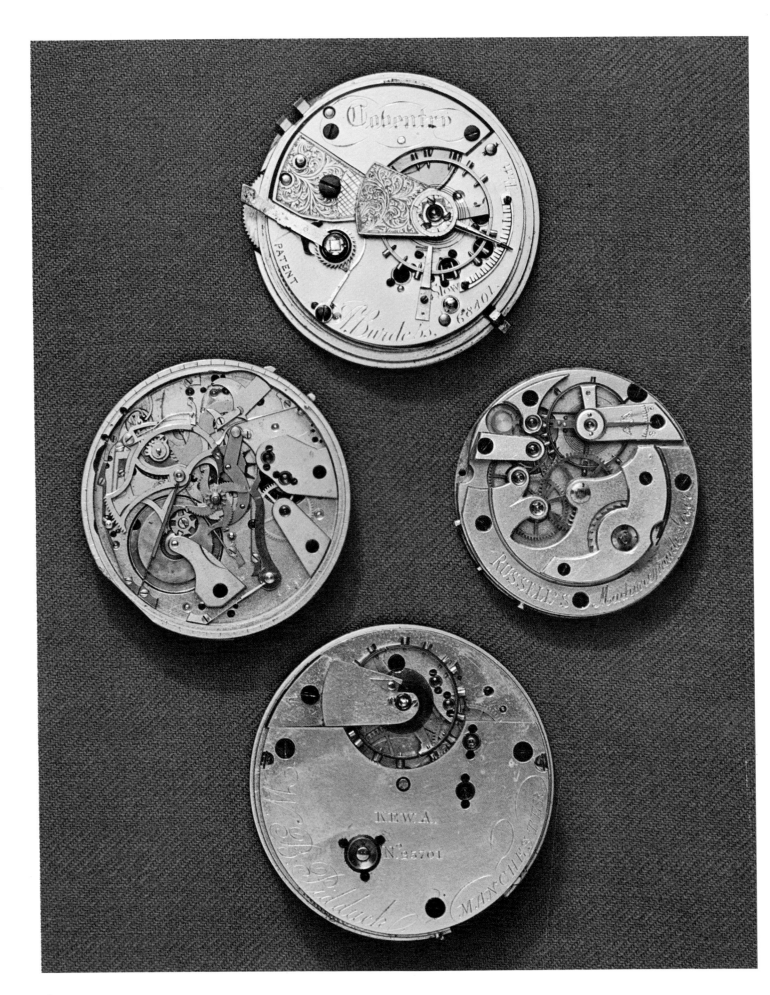

IX
WATCHES FOR THE PEOPLE

The wrist-watch is a comparative newcomer, with a history of not much more than half a century. A few were made before then, but as novelties that were generally regarded as impractical.

Probably the earliest bracelet watch still existing is one that was made for the Empress Josephine. The 18-carat gold bracelet is set with pearls and emeralds. The watch case is oval, but the dial is orthodox, and the movement is wound by a key. A duplicate bracelet, supplied at the same time, has a similar case and a dial that is almost identical, except that the 'hour hand' indicates the month, and the 'minute hand' the date. The two bracelets were made by Nitot of Paris in 1806, and presented by the Empress to her daughter-in-law, Princess Auguste-Amélie, on the Princess's marriage the following year.

However, in the account books of the Swiss firm of Jaquet-Droz and Leschot of Geneva, there is a record of a watch fixed on a bracelet that was supplied in 1790.

Certainly by 1880, another Swiss firm, Girard-Perregaux, of La Chaux-de-Fonds, and some others, were providing German naval officers in Berlin with quite small wrist-watches, about 2.5 centimetres (1 inch) across, in gold cases. These Swiss firms were encouraged to make them for the general public, but for a long time the general public was not at all enthusiastic. It was not until artillery officers in the First World War found them much more practical than pocket watches that interest grew. At the end of the war they became popular with the public. The first models were like small pocket watches held in leather cups on wrist straps.

To pick up the story, it is necessary to go back to the first attempts to make watches for everyman at much lower prices than the handmade watches of the day.

AMERICAN ENTERPRISE
The American colonists bought watches from Europe, mainly from England, and from immigrants who had set up as watchmakers.

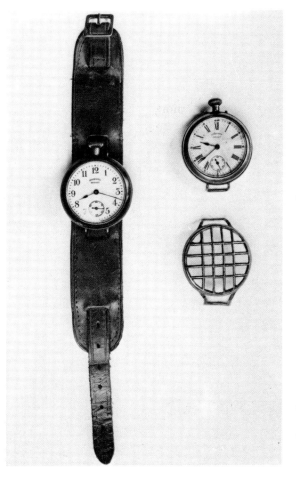

Far left: Examples of English partly machine-made pocket watches of Victorian times. The watch at the top is wound by a lever on the left instead of a winding button, and was made in Coventry. Centre right, is a watch engraved 'Russell's Machine-made Lever', emphasizing the fact, but that on the left was largely handmade by Dent, the well-known London maker. The bottom watch, although signed by a Manchester 'maker', was probably made in London. Centralized manufacture became a feature of the mass production age.

Left: Wrist-watches were late on the scene, the first ones being miniature keyless winding pocket watches fastened to wrist straps, sometimes with a guard for the glass, like this Ingersoll one of about 1914.

Right : Making jewel endstones and jewel holes out of ruby, sapphire and garnet at the Waltham Watch Works in 1885. Women, using gas flames to shellac the stones to the lathes, were employed because they could cope better, it was thought, with the fine tolerances involved.

These watches, being mainly hand worked, were expensive. Pioneering watchmakers, like the clockmakers, seeing the vast potential market for cheap timekeepers, tried to manufacture watches by machinery.

The task was more difficult. Smaller tolerances are needed in a watch, even a pocket watch, so parts had to be made more accurately to be interchangeable. This necessity became even more forcible later, when wrist-watches were made by machine. If the wheel work of a wrist-watch were magnified to the size of the machinery of an old-fashioned windmill, the accuracy of the gearing would be about the same, and a hair from the owner's head in the watch wheels would be like a log in the gearing of the mill.

It is likely that two brothers, Henry and James Pitken, of East Hartford, Connecticut, designed the first watch for manufacture by machine, although they were silversmiths. They made parts that were interchangeable, cut wheels in stacks, and used as few imported parts as possible. Their first batch of 50 watches was produced in 1838, and they are believed to have made about 400 watches.
· The man credited with solving most problems of early mass production of watches was Aaron Lufkin Dennison (1812–95), who worked for a firm of jewellers before setting himself up as a watch repairer in Boston. After further experience, he started box-making for the jewellery trade, which finally led him to the manufacture of gold watch cases.

His connections with the trade made him realize the enormous potential of the cheap watch, and he became convinced that he could design one that could be made for about a quarter of the price of the imported article. He made friends with Edward Howard, of Howard and Davis, who were clock manufacturers, and persuaded them to finance him. His first product, an 8-day watch, caused problems, so he was to make only a few. No. 1, which belonged to Edward Howard, is now in the Smithsonian Institution, Washington.

Next, Dennison tried to redesign an English watch for machine production. He was now working under a new company name, the American Horloge Company, which was changed to the Warren Manufacturing Company, after the general who had been killed at the Battle of Bunker Hill, and later still to the Boston Watch Company. The business moved to Waltham, Massachusetts, in 1854, and was soon successfully producing 30 watches a week.

The three original partners were still together in these ventures and Edward Howard looked after the finances. Many watches bore the names 'Dennison, Howard and Davis', but others were made for customers, and were named accordingly, such as 'Samuel Curtis, Roxbury'. Curtis was a mirror-maker who had helped Dennison finance the Warren company. One batch of watches was named after Fellows and Schnell of New York because they also helped the company financially.

Alas, as happened to so many pioneers, early success turned to failure, and in 1857 the company had to go into liquidation. It was bought by a New York watch importer, who initially retained Dennison as a consultant, but Dennison left after an acrimonious dispute with the directors, who tried to sack him.

The new company managed to survive, however, though it underwent several further changes in name during periods of difficulty and financial rescue, to become the Waltham Watch and Clock Company in 1923 and the Waltham Watch Company in 1925. The company made about 34 million watches until production ended in 1957. The name Waltham is still used, but as a trade name hiding a Swiss movement.

Meanwhile, Dennison had a new idea. He would have interchangeable parts made in Switzerland and assembled into movements in America. In 1863 he went to Zurich, where, with financial help from Boston, he set up the Tremont Watch Company. Escapements and trains of wheels were to be made in Switzerland and plates and spring barrels in Boston, but he fell out with the other stockholders, who were able to override him when they wanted to move the factory to Melrose. He withdrew, but the company continued to supply parts to Boston.

The new factory at Melrose failed in 1868, and Dennison returned to America to try again in Boston. There was probably never such a persistent man in the watch business. His attempts failed, so he then moved to England and sold the machinery he had bought in Boston to the English Watch Company. He settled with his family in Handsworth, Birmingham, where permanent success came to him at last.

Under the name of Dennison, Wigly and Company, he began designing machinery for making watch cases, and ran a successful business from 1874 to 1895, when he died. The business was continued by his son Franklin, then by two grandsons, Gilbert and

GAUGING HAIR SPRINGS

Andrew, under the name of the Dennison Watch Case Company, specializing in gold cases. They also founded a machine tool company and a chain-making company in Birmingham. The case-making company became one of the biggest and most successful in Europe, but eventually was sold and then went into liquidation in the 1960s.

The significance of Dennison's contribution to the mass production of watches has been contested, but there is little doubt that it was very considerable, although Edward Howard probably contributed much more than he is credited with. It seems that the three partners, Dennison, Howard and Davis, knew nothing of progress made in France by Frédéric Japy or in Switzerland by Pierre Frédéric Ingold, until they bought some of Ingold's machinery in 1845.

After Dennison had gone to Switzerland, Edward Howard continued in the original factory used by Howard and Davis, forming a partnership with the receiver to pay off the debts by making good watches. After settling the debts, he set up a company in his own name and retired at the age of 69 with a fortune.

Scores of other manufacturing companies, making watches that had to be sold

Above : Women also handled the delicate hairsprings, under large windows for daylight, and gaslights, or the rare incandescent electric lamps, by night.

for at least $10 each, were formed from about 1850 to 1875, and almost all of them ran into financial troubles. Only one survived to recent times, the Elgin National Watch Company of Chicago, which went out of business in 1964.

THE DOLLAR WATCH

The failures seemed to shift the emphasis even more forcibly on to solving the problem of producing a cheap watch that would reap the rewards of selling to an even bigger market. The ambition of every factory owner was to make a dollar watch, a watch that could be sold in the shops for a dollar. A number of them realized that the problem might be solved by making a watch much more like a cheap clock, instead of following the traditional, more sophisticated design of the handmade watch. The idea attracted into the field such clockmaking factories as Ansonia, Waterbury and E. Ingraham.

In view of this, it is a curious fact that, in order to make the first really cheap watches that were relatively accurate, designers turned to an idea that was a feature of the expensive pocket watch made by the famous Breguet, the *tourbillon* watch, which had an escapement that revolved to even out the timekeeping errors that occur with a watch in different positions. Instead of making the escapement only revolve, American designers caused the whole movement to revolve in the case.

The first to obtain a patent for such a watch was Jason R. Hopkins (*c.* 1818–1902), of Washington, in 1875. The movement revolved once every two and a half hours, and Hopkins believed it could be made for about 50 cents. After some false starts and redesign-

ing, the watch was eventually put on the market as the Auburndale rotary watch in 1877, with five jewels and a lever escapement. About 1000 were made at $10 each, but it seems that only about half of these were sold. The Auburndale Watch Company was yet another firm that could not pay its debts, and it went out of business in 1883.

The brass suppliers Benedict and Burnham, of Waterbury, were offered the watch, but turned it down as not being developed enough. The decision was obviously correct. Another watchmaker was found (by the same entrepreneur who had introduced Hopkins's watch) to design a rotary watch. He made one with a duplex escapement and a movement that rotated once in an hour. It was unjewelled and had only 56 parts, about half the usual number. Winding it was a lengthy task, however, as the mainspring was 2.7 metres (9 feet) long. It took about 150 half-turns of the winder.

Benedict and Burnham accepted this design from a company called Locke, Meritt and Buck, formed by the backers and the designer, Daniel A. Buck. A plant was set up to manufacture it. It was retailed for only $3.50, at first under Benedict and Burnham's name and then, from 1880, under the name

HIS LETTER—TO ISABELLE.

This watch, dear girl, is like yourself,
  This Waterbury "Ladies' L,"
    A dainty but a trusty elf ;
      None can deny it Is-a-belle !

Though small, you'll find it true to time,
  The moments readily to tell,
    It has no peer in any clime,
      This Waterbury "Ladies' L."

More costly gifts may grace your shrine,
  Yet none whose worth will prove so well,
    There's merit in this gift of mine ;
      You can't gainsay it Is-a-belle.

It's trifling cost will not detract,
  Because you know it Is-a-belle,
    From prizing this, although, in fact,
      It only is a "Ladies' L."

The merit locked within its case,
  Is truthful tales of time to tell,
    It has an honest, open face ;
      Can you deny it Is-a-belle ?

Then treat it kindly, and you'll find,
  No lady's time-piece can excel,
    This queen of watches, new short-wind,
      This Waterbury "Ladies' L."

The "LADIES' WATERBURY" is the neatest, prettiest, daintiest, and cheapest watch ever offered to the public. It is a perfect Ladies' Watch; Jewelled, Keyless, Stem-Set, Dust-proof; accurate and durable.

IN NICKEL CASES, 17/6.   IN HANDSOME ENGLISH HALL-MARKED SILVER CASES, 35/-

No watch leaves the factory until thoroughly tested as to its timekeeping qualities, and every one is guaranteed two years.

OF RESPECTABLE WATCHMAKERS EVERYWHERE, AND AT THE WATERBURY WATCH DEPOTS. 1885

Testimonials from Wearers all over the World. Head Office:—7 SNOW HILL, HOLBORN, LONDON, E.C.

Above : One of the Waterbury short-wind models. The twin teeth of the duplex escape wheel can be seen above the plain balance. On the right is the same watch, in a style that became classic for pocket watches of all qualities.

of the Waterbury Clock Company. The long-wind watch became famous and is now a much-sought-after collector's item. It even became a subject for cartoons, one showing an owner running beside a long fence, against which he is holding the winding button.

Some Waterbury rotaries were fitted into coloured celluloid cases, and others into more orthodox brass cases. Later versions had only 54 parts, requiring, however, 500 manufacturing operations. These were fitted into metal cases. Thousands of long-wind watches were made before production ceased in 1891.

The most famous dollar watch was made by Ingersoll, who were also the first to get the price down to a dollar. Robert H. Ingersoll and his brother, Charles H. (who was 16 years old at the time), from Delta in Michigan, in 1881 formed a partnership in New York to start a mail order business selling articles for a dollar each.

After a successful start, they concluded that they could sell cheap watches. Robert sought out a small clock with a 5 centimetres (2 inches) dial and designed a watch case for it. He backed this judgement by ordering 12,000 of the clocks from the Waterbury Clock Company, who were very dubious about the scheme. They sold for $1.50 each, with a chain. The Ingersoll watch business expanded so rapidly that, in 1894, they were able to order 500,000 for the year's supply. Two years later they sold a million watches in a year, and a watch cost only a dollar to buy.

The Ingersolls decided to give up the mail order business and to concentrate on watches, for which they coined the slogan, 'The watch that made the dollar famous'. By 1905

they had an outlet in London, selling the Crown watch for five shillings. From 1911, the British company assembled watches from parts supplied by Waterbury.

The Ingersoll-Waterbury Company survived the great depression, but, in 1944, was taken over by a Norwegian, Joakim Lehmkuhl, who had founded the United States Time Corporation, after arriving in the United States as a refugee when Hitler invaded Norway. The name of Ingersoll was retained as a trade mark until 1951, when it was superseded in the United States by Timex.

The Ingersoll Watch Company, formed in the United Kingdom in 1916, became a separate company after the Wall Street crash and continued, and still thrives, under British management. At one time the firm had a substantial share in the watch manufacturing industry in Britain, but it now employs imported movements.

The success of the Americans' efforts to make watches by machine-influenced industries in other countries, but it was some time before European watchmaking moved in the direction of mass production. Ideas for machine factories had been proposed in Europe even before they were instituted by

Right : The Yankee, a model by Ingersoll, who coined the phrase, 'The watch that made the dollar famous'.

the Americans, but European watchmakers were too conservative and had too much invested in their skills to accept change.

ENGLISH COMPETITION

England, at this time, had industries mainly in London (Clerkenwell); Liverpool, where the earliest low-priced lever watch, the rack lever, had been developed; Coventry, which lasted longest as a watchmaking centre; and Prescot in Lancashire, which supplied raw movements, *ébauches*, to the rest.

In 1858, London watchmakers formed an association, the British Horological Institute, to protect their interests. They were certainly aware of American achievements, as a report at the same time on a watch from a factory in Waltham indicates:

> The watch is a silver-bottom dome hunter, bearing the name Appleton and Tracy, Waltham, Massachusetts No. 5438. We have not been able to make so minute an examination as to be able to report upon its quality. The external appearance indicates cheapness . . . and we would seriously advise English manufacturers and workmen to lose no opportunity of preventing these articles from getting a footing

Below : Ingersoll made a watch called the 'Crown', the equivalent of their 'dollar watch' for the English market.

in colonial markets, where, whatever their quality may be found to be, they could not fail to affect to some extent the sale of English watches. Old notions and prejudices must be cast aside, and advantage taken of every means by which the standard of our own work may be maintained, while it is produced at the lowest possible price, which, considering the relative value of labour in this country and in America, ought to enable the English manufacturer to keep up a successful competition with our Transatlantic neighbours, who will let no opportunity slip of 'going a head'.

We may remark here that the specimen we have received is upon the going barrel principle, devoid of any stop work; it has sunk seconds, four pairs of holes jewelled, and a lever escapement [case dimensions follow] and is of such an appearance and style as would harmonize with the character and costume of a back-woodsman.

However, even more than the Americans, British watchmakers faced the Swiss, who were still in command of the cheap end of the European market. In 1860 there was a meeting of the Horological Institute in London, presided over by one of the country's finest makers, James Ferguson Cole, when the dangers of Swiss competition were spelled out to the watchmakers of England. Figures for Swiss exports had to be estimated, because the Swiss only provided figures for watches and clocks by the hundredweight. However, most were exported through France, which kept complete records, and the exports were estimated to be about 350,000 watches a year legally exported, plus smuggled watches, bringing the total to 500,000.

Despite their eminent makers, the Swiss had a large fringe industry noted for poor forgeries of foreign watches with the makers'

names, which were distributed by smuggling. The main speaker put it this way:

First they defraud us by imitating our names. Then their most elegant work satisfies a class to whom form and beauty are the chief requisites, irrespective of prices. These, however, are but few; it is the bulk of society we live by. The main fact is the Swiss produce a cheaper article. The English lever is a better manufacture than the Swiss cylinder; but the timekeeping qualities, irrespective of durability, are sufficient with the moderately good Swiss work.

Right : Wider distribution of watches encouraged experiment. This English watch of 1850, by Charles and Hollister, has two trains, one with a stop lever operating the independent seconds hand, and the other with a stop lever for the going train.

British production in 1858 was about 150,000 watches a year, 33,000 in gold cases and the rest in silver. Another 99,000 were imported and about 10,000 of these re-exported. About 10,740 men were engaged in manufacture, 4850 in London, 2220 in Warwickshire, 1800 in Coventry, 1160 in Liverpool, and 710 in Prescot. Those in Warwickshire were situated around Coventry.

The English watchmakers might have learned something from their own criticisms of their rivals. They complained about the cheapness of Swiss watches, then criticized the employment of unskilled female labour.

They complained about the competition offered by the small size and elegance of the Swiss ladies' watches, but would not abandon the fusee that made their own watches thick. They denied that the delicacy of Swiss work was related to female labour 'for you will see strong, big mountaineers doing some of the finest work, and English workmen can do the the most delicate work too'. But nothing was changed.

They complained that Switzerland was a cheap country, particularly in the parts where watches were made. 'Their church rates are voluntary, and there are few taxes; the people are industrious and intelligent, lovers of order and freedom, like ourselves. England is, from necessity, a dear country; it cannot afford to disband its army, and is not disposed to swamp its church.'

Education and its form of society were examined as causes of Switzerland's growing success. Cherishing of trade secrets by the British was blamed because it caused loss of opportunities for learning more from one's fellow men. In Switzerland, on the other hand, it was explained, 'owing perhaps to their political institutions, men are brought into closer contact; they have met at school, in the same *free pews* at church, and meet again at the clubs in after-life; all of which has the effect of raising the character of the people, and developing their intellect'.

In retrospect, the whole concept of watch-making was changing from that of a skilled craft industry to one that was largely de-skilled, and craftsmen would not accept the possibility of being thrown out of work by their unskilled wives and daughters.

Above : An original Ingersoll watch of 1892, showing the small clock movement adapted for a pocket watch and the dial on the left.

Frank Mercer, a chronometer maker whose life spanned the period of the demise and re-birth of the British watch and clock indus-tries, said 'the decline of our industry was largely due to the antagonistic behaviour of the old craftsmen of that period. Cole and Hammersley and others sprung these wonder-ful watches in their private houses and no new blood was coming forward. North and other makers in London had to meet the com-bined opposition of these wonderful crafts-men and I often thought all they wanted and worked for was to be in at the death.'

Watchmaking in England was still a cottage industry. In Coventry, for example, whole families worked in their own houses in ter-raced streets, fathers, mothers, sons and daughters, uncles, aunts, nephews and nieces, so, despite the objections to female labour in

factories, it certainly existed in private houses, where one room with good natural lighting was set aside as a workroom with benches and treadles. There were escapement doers (makers), jewellers (making jewel holes and endstones), half-doers (making the pivots of centre, third and fourth wheels), motion doers (finishing the motion work that gears the hour to the minute hand), engravers, dial-makers, glass-makers and others, all working at home and paid as for piecework.

Work was issued on Mondays. Small tin boxes were issued by the finishing firms, each containing the *ébauche* without pivot holes, and the various unfinished wheels. The first workman would be given a number of these and he and his family would carry out their task, setting the jewels, for example, each jewel being mounted in a brass bush. At the end of the week the tins would be returned and on the Monday issued to the next work-man, the half-doer, and so on until the move-ment was ready for final finishing. Pivots were made to fit particular jewels, so there was no interchangeability.

In the last quarter of the century, some factories had been introduced, making inter-changeable parts. Jewels still caused prob-lems. The outworker was paid a penny or one and a half pence for a jewel in ruby or sapphire, which took him about ten minutes to make. An average worker would turn out 100 to 200 a week. Jewel holes varied con-siderably in size and it was not until about 1900 that the Swiss started supplying un-finished jewels with standardized holes al-ready drilled, for two shillings a hundred. The English made the hole with a diamond-tipped tool, machining the jewel from each side. The Swiss adopted a newer method of drilling with diamond powder from one side, allowing the other to break out.

In Coventry there were about 60 so-called factories, of which that of Rotherham and Son was by far the biggest. In 1889 this was reckoned to be one of the biggest watch factories on modern lines in Europe, turning out 100 watches a day.

A major problem of the embryo factories was standardizing designs. Every watchmaker and retailer wanted small numbers of many different types. The Watch Movement Manu-facturing Company, Coventry, in the early part of the twentieth century, was prepared to make movements in batches of 500, but could only get orders in twenties and thirties.

Prescot's movement-makers decided to set up on their own making complete watches in 1890, when the Lancashire Watch Company

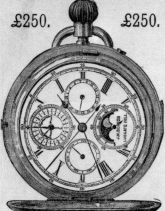   

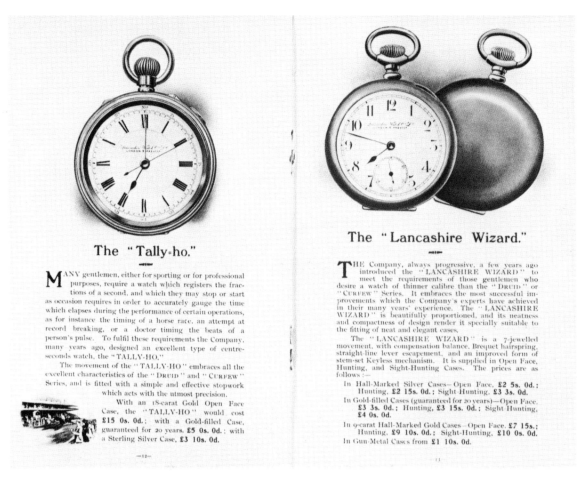

The "Tally-ho."

MANY gentlemen, either for sporting or for professional purposes, require a watch which registers the fractions of a second, and which they may stop or start as occasion requires in order to accurately gauge the time which elapses during the performance of certain operations, as for instance the timing of a horse race, an attempt at record breaking, or a doctor timing the beats of a person's pulse. To fulfil these requirements the Company, many years ago, designed an excellent type of centre-seconds watch, the "TALLY-HO."

The movement of the "TALLY-HO" embraces all the excellent characteristics of the "DRUID" and "CURFEW" Series, and is fitted with a simple and effective stopwork which acts with the utmost precision.

With an 18-carat Gold Open Face Case, the "TALLY-HO" would cost £15 0s. 0d.; with a Gold-filled Case, guaranteed for 20 years, £5 0s. 0d.; with a Sterling Silver Case, £3 10s. 0d.

—12—

The "Lancashire Wizard."

THE Company, always progressive, a few years ago introduced the "LANCASHIRE WIZARD" to meet the requirements of those gentlemen who desire a watch of thinner calibre than the "DRUID" or "CURFEW" Series. It embraces the most successful improvements which the Company's experts have achieved in their many years' experience. The "LANCASHIRE WIZARD" is beautifully proportioned, and its neatness and compactness of design render it specially suitable to the fitting of neat and elegant cases.

The "LANCASHIRE WIZARD" is a 7-jewelled movement, with compensation balance, Breguet hairspring, straight-line lever escapement, and an improved form of stem-set Keyless mechanism. It is supplied in Open Face, Hunting, and Sight-Hunting Cases. The prices are as follows :—

In Hall-Marked Silver Cases—Open Face, £2 5s. 0d.; Hunting, £2 15s. 0d.; Sight-Hunting, £3 3s. 0d.
In Gold-filled Cases (guaranteed for 20 years)—Open Face, £3 3s. 0d.; Hunting, £3 15s. 0d.; Sight-Hunting, £4 0s. 0d.
In 9-carat Hall-Marked Gold Cases—Open Face, £7 15s.; Hunting, £9 10s. 0d.; Sight-Hunting, £10 0s. 0d.
In Gun-Metal Cases from £1 10s. 0d.

13

Left : Two pages from a catalogue of the Lancashire Watch Co., Prescot, thought to have been published about 1905.

Far right : An assembly tray as used by the Lancashire Watch Co. for 15 size watches. Similar trays for ébauches and components are still in use in some factories today.

Right : The press shop at the Prescot factory of the Lancashire Watch Co.

opened a factory to accommodate nearly 600 workmen. It was reported to be a long, one-storey building with a roof shaped like the teeth of a saw, the sloping sides towards the north being of glass and admitting a good light. There was also a large room devoted to toolmaking and an engine room.

Lord Derby, who opened it, remarked:

It is only within the last few years that the art has been discovered of making them [watches] so cheaply as to be in the reach of the ordinary artisan or day labourer. They are so useful, even so necessary, that no person who can afford one will willingly do without. We have therefore opened up to us an immense market, subject no doubt to Swiss and American competition, but otherwise not likely to be much affected by the ordinary fluctuations of trade.

Lord Derby and others among the backers and management of the Lancashire Watch Company believed in free trade, but the question came up again and again of countries with protected industries selling their watches at a higher price in their own markets than abroad. It happened with watches as it continued to do with clocks. The chairman of the company complained in 1903 that, at that

moment, a man was on his way across the Atlantic with 20,000 American watches and instructions to sell them at the best price he could get for them. Importers bringing in surplus stock had to pay nothing to bring it in and were asked no questions. If he, the chairman, went to America to sell watches, he had to pay a 45 per cent tariff on them.

At that time Prescot gold watch cases contained 25 pennyweights of gold, worth £4, but 'gold' watches made abroad might contain as little as 2 pennyweights, and the complete watch could be sold at a profit for £4. The trade blamed the hallmarking laws for not letting them make inferior (cheap) gold cases to compete. Some watchmakers, it seems, would not debase their own craft, but did not mind debasing the craft of their fellow goldsmiths.

There was another way of competing: to operate from rival territory. A particularly successful concern of the time was H. Williamson, who had watch factories not only in Coventry but in Buren and La Chaux-de-Fonds, in Switzerland, producing in all just over 4000 watches a week.

After the Second World War, the British Government made a determined attempt to set up a wrist-watch manufacturing industry from scratch. After difficult negotiations the Swiss were persuaded to help, by providing automatic machinery against import quotas. Smiths, Ingersoll (by then British), and the armament manufacturers Vickers were encouraged, with financial backing of a million pounds, to set up a factory making low-priced watches in Ystradgynlais in Wales. Vickers soon withdrew and the other two companies set up the Anglo-Celtic Watch Company, and made watches successfully for a number of years.

A watch-importing firm, Newmark, started an independent factory in Croydon, Surrey, making low-priced pin-pallet watches, that is, having steel pins as the operating ends of the lever engaging the teeth of a brass escape wheel. Smiths also began making jewelled-lever watches under their own name, at a factory at Bishops Cleeve, near Cheltenham, Gloucestershire. The higher-quality jewelled lever has synthetic ruby pallets that engage with the teeth of a steel escape wheel. Smiths also inaugurated a design and research laboratory that designed the first British automatic wrist-watch to be put into production – about 30 years after it had been invented by the English watchmaker John Harwood.

Again, however, the British watch manufacturing industry died, or was allowed to when the government concluded that electronic fuses would supplant mechanical ones.

SWISS SUPREMACY

But it was the Swiss who became the world's greatest watchmakers. The turning point came with the Philadelphia Exhibition of 1876, where the 'back-woodsman' products of the American factories were on show. A

report of the exhibition was prepared for the Swiss watchmakers by Edouard Favre-Perret, who brought home an American watch, which, after he had tested it, he declared to be better in performance than 50,000 Swiss watches.

The Swiss were in a state of anxiety because they could see their industry crumbling under American competition. The authorities of the Canton of Berne even organized a competition setting several questions to be answered, two of them being: 'What is the cause of the crisis through which the watch industry is passing?' and 'What should be done about it?'

Favre-Perret gave a series of lectures insisting that the American factory methods be adopted. The warning was heeded. The message of interchangeability had been well

who was so far ahead of his time, was remembered in the early manufacturing industries for his *fraise* (strawberry), a milling cutter which was the progenitor of the milling cutters that shape all brass *ébauches* today.

At the time, however, the *ébauche*, or raw movement, was not an intricately milled plate, but plates held apart by pillars, bridges and cocks. The factory method was to stamp these out of sheet or strip brass, a method probably introduced in Europe by Japy, who supplied rough verge movements to watchmakers. Japy invented a wheel-tooth-cutting engine, a circular metal-cutting saw, various lathes for special tasks, and a tool for slitting screw heads.

The Swiss mountain watchmakers were not happy about their dependence on 20,000 movements a year which were supplied by a foreign factory, Japy's, even if it was just over the border in France, and although Japy had been apprenticed in Switzerland and had a Swiss wife. Perhaps as a result, in the year that Japy had his most successful sales to the Swiss, they opened a movement factory of their own in Fontainmelon, and a year later another in Geneva. Fontainmelon is near La Chaux-de-Fonds and Le Locle, in the Jura, which are today main watchmaking centres.

Until the end of the nineteenth century, the Swiss industry was still run very much on cottage lines, like the British, and had one major deficiency, the lack of good hairsprings, which were a speciality of the British. A springmaker in Geneva managed to develop a new spring-making process, for which he won an international prize, although the process itself was not disclosed. It was reported that even Queen Victoria enquired about it. At any rate, it improved Swiss watches and the inventor set up a company, the Société des Fabriques de Spiraux Réunies, which is still run by his descendants.

The Swiss adopted machine making of jewels. Jewel-making had originally been the prerogative of the English, and jewels were still being made by hand in Coventry in the first quarter of the twentieth century. Again, Ingold was the pioneer. He was experimenting with machine-made jewels at La Chaux-de-Fonds as early as 1823.

The keyless mechanism, essential to the modern watch because it eliminated the separate key, which is replaced by a winding button, was invented by Thomas Prest, foreman to the English maker John Arnold, in 1820, but was developed in Switzerland by Louis Audemars and Sons in 1838. The invention that became the system used today was made

Above : Some Prescot watches : top, half-hunter in a rolled gold case (c. 1905) with straight line lever Vigil movement (left), and a ladies' watch in a rolled gold case of about the same date. Bottom row, on the left the dust cap of the full plate movement in the centre, which has a dummy barrel instead of a fusee to enable it to be wound in the same direction. On the right is a silver pocket watch, hallmarked for Chester, 1901, with a similar movement, known as the Express English lever.

learned by the Swiss, who, within a few years, began developing their own machine tools.

Some of the success of both Swiss and American watch industries was owing to two pioneers, a Swiss and a Frenchman, who designed and made machine tools. The Swiss, Pierre Frédéric Ingold (1787–1878), of Bienne, worked for a time with Breguet in Paris. When Ingold was over 40 years old, he managed to interest the firm founded by the Frenchman, Frédéric Japy (1749–1813), of Beaucourt, in some of his many ingenious ideas for making watches by machine. The two companies formed the French Watchmaking Company, but it produced no tangible results for a long time, which made the backers impatient. As a result Ingold moved his headquarters to England.

Ingold and his ideas were not well received. His factory was attacked by angry craftsmen and representations were made to Parliament, with the result that his company was put out of business. He then went to the United States, where at first he was welcomed and even granted citizenship, but evidently there were forces working against him, for, shortly afterwards, he was inexplicably expelled from the country. Nevertheless, this man,

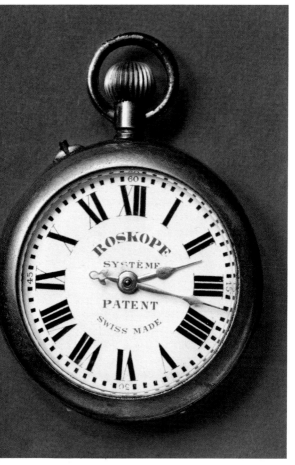

Left : G-F. Roskopf redesigned the watch movement to enable it to be made more economically. This is a later version in which the winding button also sets the hands. In earlier models they were set like clock hands.

Right : Some watches ran for eight days when wound and were called 'hebdomas watches'. The mainspring is the whole diameter of the watch.

in 1842 by Adrien Philippe, of the firm that became Patek Philippe, makers of some of the world's finest watches in the twentieth century, as indeed are the firm of Audemars Piguet.

Another invention adapted particularly successfully to mass production was the shock-absorber, first introduced by Breguet under the name *parachute*. The Incabloc, a more scientifically designed and precise shock-absorber to protect the fine pivots of the balance wheel of a wrist-watch, was invented in 1928 and became an essential feature of all better-quality watches.

One unexpected result of the Swiss adoption of factory methods was that, because they concentrated on what became better-quality watches, and eventually jewelled-lever watches, these became too expensive for a large section of the growing market. The Swiss cheap watch developed from a different source.

In the mid-nineteenth century Georges-Frédéric Roskopf (1813–89), a German watch-maker working in La Chaux-de-Fonds, set himself the task of producing pocket watches that could be sold for only 20 francs each, 'affordable by all purses'. His design was ingenious. He eliminated the centre wheel that

carries the minute hand. Instead, the minute hand was driven from the barrel. Most importantly, he simplified the escapement, replacing the cylinder of the time by a form of pin-pallet, which was much cheaper to make.

Early models were wound by a key in the old way; then he introduced simplified keyless winding, acting forwards only, with a free mainspring, not in a going barrel. In early models, the hands were set with a finger, like the hands of a clock. Roskopf intended to make the dials of strong paper or card, but could not find a supplier for the small quantities he needed. The cases were made of German silver, a copper-zinc-nickel alloy.

Roskopf was persuaded to enter one of his watches in a competition at the Universal Exhibition of Paris in 1868, where, to the surprise of those who thought his watch crude and freakish, he was awarded a bronze medal. He patented his watch in certain foreign countries, but he was unable to take out patents in Switzerland, because there were no patent laws, so anyone was allowed to make it.

He has probably not been given due credit for his achievement, for he was essentially the founder of a new, low-priced watch-making industry that was much larger in

Above : The back of a Roskopf watch. It has no centre wheel and employs steel pins instead of jewels for the escapement pallets.

volume than the traditional watch industry.

By 1936, the Swiss had passed the Americans in production, and, owing to their cheaper labour, had been able to cut prices much lower. In that year the United States, which had protected its own industries by high tariff walls, joined the Reciprocal Trade Treaty, one effect of which was to lower the duties on Swiss watches by 50 per cent. In two decades there was no longer a jewelled-lever watch industry in the United States.

By 1958, the Swiss were exporting 28 million Roskopf watches a year and Roskopf factories had been set up in several other countries, notably the United States. The design has changed, and such watches, although still called Roskopf in Switzerland, are known elsewhere as pin-pallet or pin-lever watches. They have been given a bad reputation by the traditional watchmaker and jeweller, and indeed many were poor in design and performance and wore out quickly, but many were very good indeed. In Belgium, earlier Roskopf pocket watches were supplied to employees of the railway company and called railway watches. A Swiss maker of pin-pallet wrist watches, Oris, entered some in the official trials as precision watches and won certificates entitling them to be called

chronometers. Other firms also succeeded. One, Sicura, said they would not market the watches because the certificates cost almost as much as the watches. The sector of the industry making pin-pallet watches is still much bigger in volume than the sector making jewelled-lever watches.

The mechanical watch, particularly one made in Switzerland, seemed to be the ultimate in personal timekeepers. A good one, worn on a wrist and subjected to violent movements as when playing sports, and subjected to extremes of temperature, even submerged in cold water, would still keep time to about 30 seconds a week, which is one second in 60,480, an accuracy of 99.98 per cent. There is no mechanical scientific instrument available commercially as accurate as that. There is probably no other mechanical instrument of any kind that will go for ten years and more, night and day, without stopping, without wearing out, and maintaining its accuracy.

THE ELECTRONIC WRIST-WATCH

While watchmakers were congratulating themselves about the marvels they made in such huge quantities, a tiny cloud appeared upon the horizon. Fred Lip, a prominent French watch manufacturer from Bescançon, the centre of the French trade, announced that his firm had developed an electronic wrist-watch, operated by a tiny electric power cell.

He was invited by the British Horological Institute to give a lecture on it at the Hall of the Royal Society of Arts in June 1953. So many attended that the talk had to be relayed. He was wearing such a wrist-watch, but there were problems in putting it into production, although he elicited the help of the American firm of Elgin. Meantime, in 1956, another French firm, A.T.O., who made the first transistorized pendulum clock, also showed an electronic watch.

It was the Americans who won the race, however, when the Hamilton Watch Company, which had an interesting history of innovation and was the only firm that managed to mass produce marine chronometers during the Second World War, announced in late 1956 that they actually had electric wrist-watches in production, and that they would be selling them, in gold cases, on 3 January 1957. It was an exceptionally well-kept secret.

In the electric watch, the mainspring barrel was replaced by a power cell that was a little larger but gave enough energy to run the watch for a year. A small coil of extremely

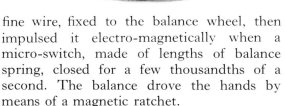

fine wire, fixed to the balance wheel, then impulsed it electro-magnetically when a micro-switch, made of lengths of balance spring, closed for a few thousandths of a second. The balance drove the hands by means of a magnetic ratchet.

Other electric watches followed, from France, Germany and Switzerland. The Lip watch incorporated a transistor for switching, which was why it was called an electronic watch, and the Swiss developed a transistorized system that was particularly effective.

In 1960 there was another shock for the makers of mechanical watches. Max Hetzel, a Swiss electronics engineer who had worked in Swiss factories on rate-recorders that controlled the timekeeping tolerances of products, invented a watch that used a tuning fork as an oscillator, combined with an oscillating electronic circuit. He could find no one in Switzerland interested in such an advanced idea, so he went to America. There the first space programme was under way and there was a need for a small, accurate time switch, that could operate sequences in artificial satellites or be left on the moon and could be switched off by a signal or automatically to avoid cluttering up space with unnecessary signals. The Bulova Watch Company, which had a contract with N.A.S.A., took up Hetzel's scheme and developed it not only as a time switch, but also as a wrist-watch movement, called the Accutron.

A tuning fork about 2.5 centimetres (1 inch) long was made to vibrate at 360 Hz by magnets and electro-magnetic coils (transducers) in an oscillating circuit powered by a cell that lasted for at least a year. The vibration of one of the tines of the fork drove the hands

Left : One of the first Hamilton electric movements, introduced in 1956. The battery at the top has been removed. The balance operated a mechanical switch.

by a tiny ratchet arrangement, the ratchet wheel being 2.5 millimetres (1/10 inches) in diameter and with 300 teeth. This electronic watch had timekeeping accuracy built in, which enabled the makers, for the first time in history, to guarantee the timekeeping during the life of the watch, to a minute a month.

The Swiss were slow to react to this major change in technology, perhaps because they had so much invested in the mechanical watch industry and had an underdeveloped electronics industry, but they did set up various organizations to study electronic watches, the main one being the Centre Electronique Horloger (C.E.H.). The race to produce a true electronic wrist-watch was on; a race to miniaturize the already developed quartz crystal clock to wrist-watch size.

In 1967 the C.E.H. announced a quartz wrist-watch, an amazing feat of miniaturization. Moreover, it was about ten times as accurate as a conventional watch. Some were entered for official timekeeping trials and swept the board. It seemed that the Swiss were ahead again, but it was not so. The Japanese had also been working on the problem and entered their own watches in the next Swiss timekeeping trials. The Swiss

Right : The first major step towards a true electronic watch was the introduction of the electronically maintained tuning fork watch by Bulova. The inventor, Max Hetzel, returned to his native Switzerland where it was known as the 'sonic watch', and developed it a little further.

for digital watches that have recurred usually two or three times a century almost since the watch was invented. This provided the visual novelty value needed by an owner to indicate to everyone around him that he owned one of the new electronic watches. Every watch manufacturer suddenly wanted to have a range of quartz digital watches, and most of them succeeded, but the Swiss had to depend largely on the Americans.

The Japanese were as swift, moving into mass production of electronic watches, but based on their own watch manufacturing industry, and not the electronics industry of a rival commercial nation. The Japanese watch industry was a phenomenon of the years following the Second World War. It had blossomed like the camera industry and had gained a high reputation by strict quality control of exported goods. Part of its success was owing to efficient centralization free from labour troubles. The introduction of automation was encouraged by both management and employees, instead of being resisted, as in older industrial societies. They were therefore well placed to develop the quartz watch and sell it throughout the world.

The advent of the quartz watch and the Japanese challenge had a severe effect on the whole Swiss industry, which went into recession. In order to meet the new challenges the 550 or so watch manufacturers and finishers, and also the huge *ébauches* manufacturing concern, were forced during the 1960s into massive reorganization into several large corporations. It must have been a familiar pattern to the many families who remained in the Swiss industry since the threat of the American factories in the nineteenth century.

The real challenge was in the technology of making watches. It was blurred somewhat by the broad assumption of those who did not know their horological history that the digital dial itself was a breakthrough in the way of telling time, and that this was the real innovation. The Japanese seem to have evaluated the situation more accurately, as they planned for a major proportion of quartz production to be with orthodox dials and hands, now renamed analogue displays.

The American electronics manufacturers soon realized that the economic strength of the Swiss was now not in their technology but in the acceptance of their brand names. The Swiss had little manufacturing muscle in the new technologies, and Americans, with the help of another technology at which they excelled, advertising, could establish brand names also. So the Americans devised brand

cancelled the trials. The Japanese firm of Seiko was the first in the world to put quartz crystal wrist-watches on the market.

Some felt that the true electronic timekeeper must have no moving parts. The C.E.H. and Seiko watches had hands. First to put into production a solid state watch without mechanical moving parts, except for a button, was again the American firm of Hamilton. On their Pulsar watch the time was shown in digital form with bright dotted figures lighting up on a dark screen when a button was pressed. If pressure on the button was maintained, seconds appeared below the hours and minutes and counted off the time.

It requires a large and costly organization to manufacture micro-circuits and microprocessors. In America the main ones were Optel, Hughes Aircraft, Fairchild and Texas Instruments, all in the computer business. Soon the principal Swiss watch manufacturers, afraid of losing much of their traditional business, were making contracts with American firms to supply quartz crystals, chips, or complete electronic modules, while still continuing their own research.

The boom came indeed. The public imagination was truly caught by the electronic watch. It generated one of the cyclic fashions

Far left : Making electrical joints in a modern quartz electronic watch is on such a miniature scale, it has to be done automatically with the operator checking it by a magnified image on a TV screen. This operation is in the Japanese Seiko factory.

Left : In some modern quartz crystal watches, such as the Omega chronoquartz, both analogue and digital displays are incorporated, as the first is best for time of day and the second for measuring time intervals.

names and supported them by massive promotions all over the world, with considerable damage to Swiss sales. However, they made a fundamental mistake.

The pocket calculator had been developed slightly before the quartz watch. It was spawned and grew rapidly through many mutations in an atmosphere almost of technical anarchy. Manufacturers, particularly in America, developed miniature chips containing complex electronic circuits, and sold this raw technology to the public, who were bemused by the marvel of it all. As soon as the latest miraculous calculator (and it was almost a miracle) appeared at an attractive price, an even more miraculous one would be announced at an even lower price.

Now the Americans marketed the quartz watch in the same way as the pocket calculator, assuming that each new model would go out of fashion and be abandoned in the same way, every time a newer and more exciting and cheaper version appeared. However, they were making a basic error. They failed to take into account the fact that the attitude of the watch-owner, encouraged over centuries, is that a watch should go accurately for ever and never wear out or become out of fashion. This attitude was the saviour of the Swiss watch industry, because it gave the Swiss breathing space.

There is a high rejection rate in chip manufacture and it is costly to set up quality controls for every chip, so manufacturers adopted the practice of letting the public do their quality control for them; they offered to replace any faulty instrument by return post without charge over a specified time. The faulty computers could then be examined

after a period of use and design and production methods be modified if necessary. With some of the early electronic watches, marketed on the same principle, every single watch was returned.

In this situation, it is usually the retailer who is blamed, rather than the maker, so retailers began to refuse supplies. Manufacturers had to go more and more to direct outlets, where price was the only consideration and advice and service non-existent. By the 1970s some of the American concerns had had enough, as had several Swiss companies who followed the same theories, and one successful English electronics company, Sinclair, which, despite the world-wide success of their calculators, stumbled in the same way over their electronic watch and had to seek state financial aid.

In the meantime the patient restructuring of their industry by the Swiss was beginning to pay off, so that by the later 1970s they were making a strong comeback, with their main competitors the Japanese. The huge Ebauches S.A. manufacturing complex in Switzerland had founded various electronics factories to complement its many mechanical watch factories. One such factory was turning out 10,000 quartz watches a day in 1977.

Above: Micro-circuitry is so small today, that complex computers can be incorporated in the watch, as in this Hewlett-Packard American version. It is also possible to store telephone numbers and birthday dates in some modern quartz watches.

Japanese production had expanded too, in both quartz and mechanical watches.

The modern quartz wrist-watch will keep time to a minute a year, if it has a good module. The quartz crystal itself may be a tiny bar, machined out of a larger piece, which flexes when it vibrates, or in the form of a tiny, flat fork, which is produced by a photo-etching process, and vibrates like a tuning fork. The quartz, which was formerly cut out of rock crystal from Brazil, is now made synthetically in a factory.

The crystal has to be delicately mounted in order to be free to vibrate in the correct mode, and it is contained in a minute vacuum chamber to reduce the effects of temperature on its rate. It is the most vulnerable part of a quartz watch and has to be shock-proofed, like the balance and spring of a mechanical watch.

The usual frequency adopted for the crystal is 32,768. This compares with the 5 to 10 Hz of a balance and spring and the 300 to 400 Hz of a tuning fork or sonic watch. The frequency of 32,768 is chosen because it is a multiple of 2 and represents a convenient size of quartz to fit in a watch case. It is 2^{15}, which means that division 16 times by 2 gives a frequency of 0.5 Hz, or one beat a

second, like a dead-beat seconds hand or digital display. The dividers work in this way, by binary counting, which only needs two states, on or off, conducting or not conducting, a blip or no blip.

A tiny integrated circuit provides an oscillator to vibrate the quartz crystal, and be controlled by it. The circuit will also provide a divider chain to break down the high frequency to one pulse a second. In an analogue watch, this drives a stepping motor (a tiny electric motor that oscillates or rotates in jumps) to operate normal hour, minute and second hands. In a digital watch, the integrated circuit is more complex. The chip will also contain more dividers that provide pulses every minute, hour and day, as well as a decoder that translates these into figures on a display.

Although stepping motors are usually developed especially for quartz analogue watches, the electronic components have become so small that it is practicable to use a traditional oscillator. For example, Timex in their quartz watch use a normal balance and spring monitored by a quartz module, and Bulova used a tuning fork overrriden by quartz.

The first solid state watch, the Hamilton Pulsar, had a display made of early light emitting diodes (l.e.d.). An l.e.d. is a form of transistor that emits light when conducting. Those for the Pulsar were handmade and cost up to $10 a digit, each digit being made of as many as 20 or more diodes. Eventually monolithic (one-piece) l.e.d. displays were developed, which reduced the cost.

Such diodes are not ideal for watches because they use a comparatively high current, which means the watch cannot show the time continuously. The watch is blind until the owner presses a button, a major disadvantage when he is hurrying to catch a train with baggage in both hands. Normally the display shows red numerals, or letters for a calendar, against a black background, so this alphanumeric display is better in darkness.

In the 1970s, another form of display was developed, called the liquid crystal (l.c.d.). The liquid, imprisoned between very thin sheets of glass, was either transparent or opaque according to whether an electric charge was or was not applied to it. When it was transparent, a mirrored surface behind it became visible by reflection. It used so little current that it could be used continuously, but the display depended on light falling on it to be seen. It had the opposite characteristics from the l.e.d.; it was good in a bright light but no good in darkness. This prompted some manufacturers to incorporate a miniature light for use in the dark.

Another disadvantage of the earlier l.c.d. was that it had a relatively short life of about two years. The principle was much improved in the later 1970s, however, with the introduction of what is called the field effect. This kind of l.c.d. has a long life and is clearer because it shows black figures or letters against a light background.

An l.c.d. is relatively slow to respond, so that displays can be seen to dim and brighten as they change, which is not important when indicating seconds and longer intervals as time of day. It is not satisfactory, however, for accurate timing, so the best timers still have l.e.d. displays; in fact some specialized watches, electronic chronographs incorporating time of day and timing facilities, have both kinds of display.

The makers of electronic watches have made amazing strides. Some of their designs have solar cells on the case that recharge the power cell from a brief exposure to natural or artificial light. Others 'are combined with wrist calculators, operated by the tip of a ballpoint pen. Most give days and dates and some have perpetual calendars programmed to accommodate leap years. Others are being designed with memories for storing friends' birthdays or telephone numbers. But their accuracy is their most remarkable virtue. Indeed, almost any quartz watch that can be bought from a jeweller's shop today for a week's wages or less is more accurate than the finest precision clocks of 70 years ago.

THE FUTURE OF THE MECHANICAL WATCH

It is often asked if the mechanical watch has a future. The answer is yes. Production is still many times that of the quartz watch and is likely to be so for some years. A mechanical watch will go for years without cost and if it is automatic does not even need winding.

There have been small developments in mechanical watches to improve timekeeping. One is the fast-beat movement. Such a watch ticks at twice or more the rate of a normal watch, which reduces timekeeping errors. There have also been major improvements in rationalizing designs to reduce numbers and varieties of parts in order to simplify production and repair. Some of the Japanese assembly lines for mechanical watches are totally automated, even the assembly of all the miniature parts into a going watch.

Some concerns have tried to produce a

Right : Progress on mechanical watch production still continues, side by side with electronic watch production, in the main producing countries, particularly Switzerland and Japan, where this completely automated production line belongs to Seiko.

cheap watch with a good timekeeping performance that is long-lasting and never needs servicing. The most successful to date has been Tissot, who, in 1970, announced a watch movement made entirely of plastics. There were only 52 parts instead of over 90 in a comparable metal watch. The *ébauche*, which normally takes 40 operations to make, was moulded in one. Most important, all the plastics used were self-lubricating, which did away with the need for oil, the bane of all mechanical watches because, after a time, it forms either a grinding paste with quartz dust out of the air, or a hard coat like varnish.

Despite the immense strides in electronics, however, mechanical watches are likely to remain the personal timekeepers in developing countries for some time. They can be produced by a developing country without dependence on the high technology and foreign exchange needed to buy and use electronic modules; they do not need a network of service stations to supply power cells every year and do not incur the increasing cost of such cells; they need no attention for years in places where it would be impossible to find a watchmaker or electronic specialist; and their mechanisms can be understood and repaired by an interested mechanic.

X
THE SCIENCE OF TIME

The belief that the future can be foretold by the alignment of certain planets with the Earth and sun has persisted for nearly 5000 years, since the Babylonians developed astrology as a system. Kings of Babylonia and Assyria had their royal astrologers, as did the kings of Egypt, the Roman emperors and the popes of the Middle Ages. Queen Elizabeth I had an astrologer, John Dee, while the French kings were much influenced by the predictions of Nostradamus.

Astrology was divided into natural astrology, concerned with the motions of the heavenly bodies, and judicial astrology, which interpreted these motions in relation to terrestrial life and events. In the late seventeenth century, when Sir Isaac Newton disproved the old superstitions, natural astrology was separated in the West to become astronomy. In the West, judicial astrology lives on mainly as entertainment, but in some countries it is still a living force: the King of Nepal's coronation was delayed three years until 1975 because the signs were inauspicious.

To the wise men of Babylonia and Egypt nearly 2000 years ago, the world was an island covered by a dome, the firmament, where gods carried lamps in the night. The whole was surrounded by water which occasionally seeped through the dome above. Priests and astrologers built towers from which to study the lamps or stars, noting their positions and making maps.

They discovered that the stars were in fixed places in the firmament and entered the night from the east and left by the west in never-changing succession. They also became aware that the double-faced god, Janus, keeper of the gates, let through some vagabonds that

wandered about, but never left a narrow path in the heavens at an angle of about 23°.

The astrologers named this path the zodiac, and divided it into 12 sections which they named after the nearby constellations of fixed stars: Capricorn, Aquarius, Pisces, Aries, Taurus, Gemini, Cancer, Leo, Virgo, Libra, Scorpio and Sagittarius. This enabled them to locate the wandering stars (planets).

Far left: Astronomy and horology have always been closely linked. This picture is from Selenographia *by the astronomer Helvelius.*

Left: Astronomy is the scientific element of astrology, of which the most famous practitioner was Michel Nostrodamus, seen here.

In about 500 BC Pythagoras discovered the relationship of the length of a musical string to the note it produced. He built up a complex numerical pattern in natural events and founded a brotherhood to worship this new order. He pronounced that the planets (which included the sun and the moon) were fixed to transparent spheres around the Earth and the different rates of their revolutions caused different notes of music. For example, the interval between the Earth and the moon was a tone, and that between the Earth and Mercury was half a tone. Only Pythagoras could hear the great celestial lyre playing the music of the spheres.

Although confused by magic, the conviction of Pythagoras that numbers were the key to the universe, and indeed to all scientific knowledge, was a big step in man's thinking. His discovery of irrational numbers, that were both odd and even at the same time, caused the collapse of the Brotherhood. (An irrational number is a whole number such as the root of two that cannot be expressed as a ratio of other whole numbers.) Unspeakable numbers, as they were called, were kept secret until revealed by a pupil. As a result, it is said, Pythagoras was put to death.

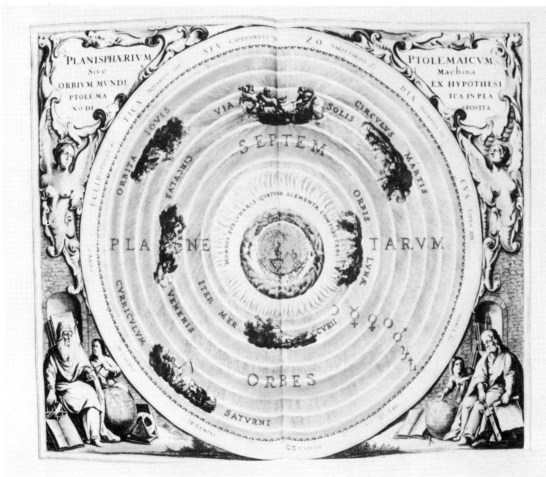

Left : Measurement of the day by the passage of the sun made it seem obvious that the Earth was the centre of the universe. These two pictures illustrate this belief, attributed to the great astronomer Ptolemy. That above shows Aristotle's four elements – earth, air, water, and fire – surrounded by the spheres of the planets and sphere of the fixed stars (c. 1523). That on the left also shows the elements as they were believed to be and is dated much later, c. 1708.

In the fourth century BC, Aristotle suggested that the Earth might not be flat. The idea of an oyster-shaped world gave way to that of a round one floating in air. The Greeks' universe was a series of planets circling the Earth and keeping to the zodiacal belt. In order, they were the moon, Mercury, Venus, the sun, Mars, Jupiter and Saturn. Although to the astrologers it was obvious that all of them circled the Earth, their capricious movements were incomprehensible. One or another of the planets would slow down, stop, drift backwards, and then wander off in the original direction again.

Aristarchus suggested in the third century BC that the explanation might be that the Earth was not the centre of the universe, but this idea seemed highly unlikely. The first natural astrologer or astronomer of giant stature was Ptolemy of Alexandria, who lived in the second century AD. He made his star maps with Earth as the centre. He plotted movements of some planets with precision, to find that they travelled in a series of loops like an aeroplane continuously looping the loop.

Ptolemy's sky maps were so accurate – he was the first to divide the degree into minutes

Right : Ptolemy using a quadrant in an illustration of 1508. He was an intellectual giant whose ideas of the universe and maps of the known world persisted for centuries with both thinkers and sailors.

*Right : An astrolabe
made by Georg
Hartmann of
Nuremberg and dated
1548.*

*Below : The astrolabe,
developed by the Arabs,
was a technical triumph
of the ancient world.
Here Arabs are using
astrolabes in a
manuscript, 'Les
Maqamat d'Abou
Mohammed al-Qasim
ibn Ali al-Hariri' of
the thirteenth century.*

and seconds – that 1200 years later Columbus and Vasco da Gama were able to navigate across vast unknown seas with them as a navigational guide.

There was in instinctive or indoctrinated belief common to all early astrologers that their gods made the world perfect in man's eyes. The universe had to be constructed on the perfect principles of Euclid's geometry and Pythagorean numbers. Unfortunately, observation did not bear out this truth. Even Ptolemy's epicyclic loops were not circular but awkward egg shapes. To illustrate the actual motions of the sun, moon and planets circling round the Earth needed 39 wheels plus one representing all the fixed stars on a rotating firmament beyond the planets.

From early times, clockmakers and mechanicians made models of the universe for the thinkers and astrologers. We know that in the third century BC Archimedes had a planetarium, a machine to show the motions of the planets, because Cicero commented upon it, admiring the fact that one revolution of the machine showed the various different motions of the planets represented. When Syracuse fell to the Romans, two globes by Archimedes were taken as spoils to Rome. It is thought that one was a star map, and the other a form of planetarium worked by hand or by water. Cicero described the planetarium as a species of globe in which the sun and moon were made to revolve with 'five of those stars, which have been called travellers, and as it were wanderers. . .'. It demonstrated eclipses correctly, he pointed out.

Archimedes' moving sphere was so famous in antiquity that it may have been the inspiration for a series of water-driven celestial clocks and astronomical instruments which culminated in the great astronomical clock of Su Sung.

The Greeks knew how to project a star map on a flat surface as well as on a sphere. Ptolemy used this method so that his flat maps were accurate. This technique resulted in the invention of the so-called anaphoric clock. A Greek astronomer had a disc engraved with an image of the sun which was made to revolve by water-power behind a fixed grid. Horizontal lines of the grid showed the altitude of the sun and vertical ones its azimuth or angular distance along the horizon.

Dr Derek Price, of Christ's College, Cambridge, has pointed out that this astronomer's model was truly the first clock dial and set the standard for hands turning clockwise. In the northern hemisphere, looking south, the sun moves clockwise. A rotating dial and fixed pointer were common on the earliest clocks before the moving pointer and the fixed dial became commonplace.

The anaphoric clock with the fixed stars engraved on the disc became the planisphere, a star map rotated to show star time and the movements of the constellations in the heavens. Planispheres were incorporated in many later astronomical clocks. They showed the appearance of the heavens at a particular time and place.

Geared astronomical calculating machines of considerable complication were made in Hellenic times, to judge by the remains of the earliest one known, the *Antikytheria*, which was recovered from a sunken treasure ship in the Mediterranean in 1901. Some of the wheels were egg-shaped, like the Ptolemaic

Left : An armillary
sphere made on the
Earth-centred
Ptolemaic system by
Adam Heroldt of
Rome in 1648. Such
spheres were used to
teach astronomical
ideas.

Right : An anonymous
sixteenth-century
engraving of the
astronomer Copernicus,
holding a model of his
sun-centred universe.

orbit of Mercury. The three-dimensional geometry that made the anaphoric clock possible also gave birth to the astrolabe in the Islamic world.

The Chinese and the Arabs undoubtedly made many globes and spheres showing the motions of the heavenly bodies, although none have survived. A globe is taken to mean a solid and a sphere to be a central ball (the Earth), with rings around it depicting features of the universe. The most common is the armillary sphere, which can also be used for timetelling. It looks like a skeleton globe with metal rings for the tropics, and equator.

When Ptolemy's theories were transferred to the West through the translations by the Hebrew communities in Spain and southern France, and by a Benedictine monastery in the Pyrenees, they awoke a desire among the learned communities of Europe for machines to demonstrate the operation of the universe. But the man who had most influence on astronomy from the sixteenth century on, cared little for such machines.

COPERNICUS

Nicholaus Koppernigk (1473–1543) lived in an area of east Prussia that was often a battle-ground. Much of his life was spent in a tower of a castle where he acted as his uncle's physician. He had a platform built high up on a wall, and from here he observed the stars with crude instruments he made himself, following Ptolemy's instructions written 1300 years previously. He could have had much more accurate ones made in Nuremberg or Augsburg, but he became more interested in the mathematics of astronomy. He is usually known by his Latin name of Copernicus.

His observations dwindled as he spent his time writing two manuscripts, one a translation into Latin of some Greek epistles, and the other his *Commentaries*, a brief outline of his views of the universe. The epistles were printed. The outline of the universe was circulated in manuscript form to some thinkers of the time, and then lay in a drawer. Later, after he had become a canon of the Church, he wrote his *Book of Revolutions*, also about the universe, but that too was circulated and never printed.

He revived the idea of a sun-centred universe, but it raised hardly a ripple of interest among scholars. Even his refusal to attend a conference of astrologers and mathematicians on the reform of the calendar, on the grounds that the motions of the sun and moon were not precisely known, was hardly noticed.

Above : A planetarium by James Ferguson, based on the Copernican system.

Copernicus never published his writings on the heliocentric system, and it was left to his disciple, Rheticus, to publish a book under his own name. The book was brought to the bedside of Copernicus a few hours before he died. A thousand copies were printed in Nuremberg in 1543, but few were sold. The only scientific work published by Copernicus himself was a set of tables based on inaccurate and out-of-date observations. They were no improvement on Ptolemy's tables as far as navigation was concerned, and were soon forgotten. Even Copernicus was ashamed of them.

One of the Copernican hypotheses was that the heavenly bodies do not move round the same centre; the Earth is only the centre of the moon's orbit; the sun is the centre of the planetary system and the universe; and the apparent rotation daily of the fixed stars is due to the Earth's rotation on its axis. In other words, the paths of the planets were simpler if looked at from the sun, but they were still circular.

It was a reasonable step, but unfortunately Copernicus made it more complicated than Ptolemy's system by retaining circular orbits and unwittingly making the planets revolve around the centre of the Earth's orbit instead of around the sun. To draw or make a model required 48 wheels instead of the 40 of Ptolemy. It did explain, however, how the planets apparently stop and move backwards for a time before resuming their forward travel. The impact on clockmakers at the time was nil. They continued to make globes and spheres with the Earth in the centre for many centuries.

Below : A diagram of the sun-centred universe attributed to Copernicus, published in 1660.

From about 1500 to 1700, great numbers of globes and spheres were built in Europe, many driven by springs so that they worked and showed the time. There are probably more sixteenth-century globes and spheres in existence today than there are clocks and watches of the same time. Some were used solely for natural astrology; others have quite separate rings for the 12 houses of the heavens and were used as well for astrological prediction. To the astrologer, the houses are like the hour numerals of a clock. They are stationary in the ecliptic, the apparent path of the sun around the Earth. The signs of the zodiac are like 12 clock hour hands that move around the houses.

The Ptolemaic system persisted for centuries in globes and spheres. It was the working system for astronomy, for navigation and for astrology. Astrology is still based on a universe with the Earth at the centre, and always will be, which is one reason why it is looked upon with derision by most scientists.

JOHANNES KEPLER AND TYCHO BRAHE

The man who made the next dramatic move was a strange mathematician-astronomer from southern Germany by the name of

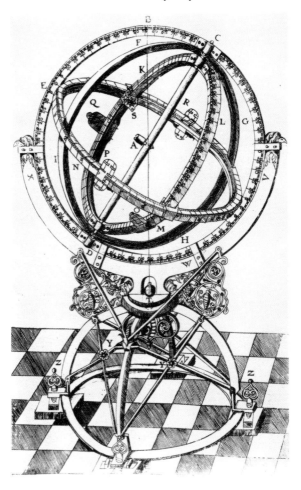

Johannes Kepler (1571–1630). He constantly suffered from various illnesses and he was also extremely myopic; he had to devise what became the modern astronomical telescope in order to see the stars. He too joined the clergy and taught mathematics at a university where one of his tasks was the preparation of the annual calendar of astrological forecasts. Astrology eventually became his profession, and he ended his career as Court Astrologer to the Duke of Wallenstein. Although he called astrology a 'dreadful superstition', he wrote serious treatises on it.

In 1594, at the age of 23, Kepler had a revelation in which he saw the 'true pattern' of the universe, based on the theories of Pythagoras and Copernicus. Certain beliefs of the Pythagorean Brotherhood were based on the five perfect solids – the four-sided pyramid,

Above: The astronomer Johannes Kepler, from an illustration dated 1870.

Left: The astronomer Tycho Brahe's equatorial armillary sphere with a moveable equatorial ring (c. 1580).

Right : Tycho Brahe's mural quadrant and two clocks, shown in his Astronomiae Instauratae Mechanica *of 1602.*

the cube, the octahedron, the dodecahedron and the icositetrahedron – because each could be inscribed in a sphere. Here was the secret of the universe, the mathematical distances between planets.

Remarkably, a dodecahedron fitted between the orbits of Earth and Mars, an icositetrahedron between Earth and Venus and an octahedron between Venus and Mercury. But, even on the poor observational information available to Kepler, the theory was not proven, and for a time he returned to the musical spheres of Pythagoras. He also resolved to study under the great Danish astronomer Tycho Brahe, who had a collection of clocks, globes and spheres and other instruments that were the wonder of the civilized world, and whose measurements of the heavenly bodies were by far the most accurate of the time.

Tycho Brahe (1546–1601) had an unorthodox upbringing. He was kidnapped from his cradle by a rich uncle and taken from his home in Helsingborg Castle (which Shakespeare made the home of Hamlet, Prince of Denmark). While a student, he lost the bridge of his nose in a sword fight at midnight and replaced it by a false bridge made of wax and a gold and silver alloy. Seeing an eclipse of the sun and reading Ptolemy at the age of 14 set him on the path of astronomical observation from which he never deviated for the rest of his life. Although he was rich and an aristocrat, he had no interest in frivolous pastimes. A wonderful observatory was built for him by King Frederick II at Uranienborg on the Isle of Hven, which now belongs to Sweden, and he had four superb clocks and globes, one with a wheel of 1200 teeth, made by Jobst Burgi, and others by different makers, including Habrecht. Two were monumental astronomical clocks.

In 1596, Tycho joined the famous team of scholars and thinkers that the melancholic Emperor Rudolph II, whose main interests were astrology and alchemy, gathered around him in Prague. He took his four Burgi clocks with him, because he had fallen out with the Hven islanders and lost his court support in Denmark. The Emperor persuaded Burgi himself to join the other celebrated *organopeos*, as mechanics, instrument-makers and clockmakers were called at the court. Perhaps it was on the recommendation of Tycho, who had received a letter from Kassel, where Burgi was the court clockmaker. The letter from the Landgraf there said: 'Recently the accurate longitude of Orion, Canis Minor and Major have been observed by his [Burgi's] clocks,

indicating minutes and seconds with such accuracy, that between two culminations, they deviate less than one minute.'

Jobst Burgi (1552–1632) was a Swiss who is thought to have assisted Isaac and Josias Habrecht in the construction of the second famous monumental astronomical clock in Strasbourg cathedral. After learning to design and construct astronomical clockwork at the court of Kassel, and having made improvements in the timekeeping of clocks, he spent the second half of his life at the Emperor Rudolph's court with Tycho Brahe and then Kepler.

It was very unusual at the time for a clock to show minutes, and unknown for it to do so accurately. The pendulum clock had not been invented. It may have been Burgi who modified the verge escapement, replacing the

Below: Jobst Burgi's rock crystal clock, made in Prague in about 1615, probably the finest he made. It has a cross-beat escapement. Burgi was also a remarkable mathematician and the first to compile and use a form of logarithms.

but also the principles of instrument optics.

The first two laws of Kepler were that the planets travelled in ellipses with the sun at one focus, and that they did not travel at uniform speeds but at speeds that swept equal areas of the ellipse in equal times. This eventually gave the astronomical clockmakers something to think about.

His third law, also related to orbital speeds, came later and was buried in some non-sensical astrology in his book *Harmonices Mundi (Harmonies of the World)*.

Kepler also provided the world with a new entertainment. He was the first to write science fiction in a story that was published after his death about a young boy from Thule whose mother sold him to a sea captain. The boy became a pupil of Tycho Brahe and was rocketed to the moon. The book even contains the concept of zero gravity.

Tycho's death left Kepler more problems than he could have forecast. A swashbuckling son-in-law promptly sold Tycho's priceless collection of clocks and instruments to the Emperor, whose Treasury did not pay for them. Consequently, they were locked away out of Kepler's reach until, sadly for posterity, they rusted and deteriorated. Kepler did take all Tycho's papers, and admitted to stealing them and refusing to hand them over to the son-in-law, who proposed that he would cause no more trouble if Kepler published his works under their joint names!

Magnificent as were the globes and spheres made for the astronomers, they were only as good as their accuracy of gearing, and the clocks were only as good as their oscillators. Observing astronomers more and more concerned themselves with accuracy in time-keeping, realizing that their mathematics might produce results where the pragmatic approach of the clockmaker had failed. The stars they watched appeared to provide the notion of time, which could only be measured by the apparent motion of the same stars, as diamond was the means of cutting diamond.

Left : The cross-beat escapement in the second experimental clock made by Burgi. The escapement was the first attempt to improve the accuracy of the verge and balance. Two vertical balances swing in opposition. The clock was made before 1600.

Right : A portrait of Galileo by Ottavio Leoni of Rome (1587–1630).

balance bar or wheel by superimposed centrally pivoted arms with weights on each end that oscillated in opposite directions. This has been renamed the cross-beat since it was investigated by Dr H. von Bertele in 1954.

Tycho was astrologer to Rudolph II, but not because he was a mystic like Kepler. According to the account by Arthur Koestler in his history of the early astronomers, Tycho's domineering character only allowed stark superstition. His individual contribution to astronomy was a remarkable new, detailed and accurate map of the heavens. As important, he supplied to Kepler (sometimes unwillingly) a mass of accurate data that enabled Kepler to turn the speculations of Copernicus into mathematical theory.

After two years of correspondence, Kepler went to Prague at Tycho's invitation, and they met for the first time on 4 February 1600, an arrogant Dane of 53 and an eccentric German of 29. The association was punctuated with acrimonious rows, and within 18 months Tycho was dead.

Two days after Tycho was buried, Johannes Kepler was appointed Imperial Mathematicus in his place, and from his cranky start with perfect solids, eventually founded not only the science of physical astronomy

GALILEO GALILEI

Galileo Galilei (1564–1642), the son of an Italian intellectual, became interested in accurate timekeeping while he was still at university. By a simple but remarkable feat of observation, in 1582 he discovered that the time in which a pendulum swings depends only on its length, not on its amplitude – how widely it swings – or on the weight of its bob.

Three years before he met Tycho, Kepler sent to Galileo his work *Cosmic Mystery*, setting out his views on the Copernican

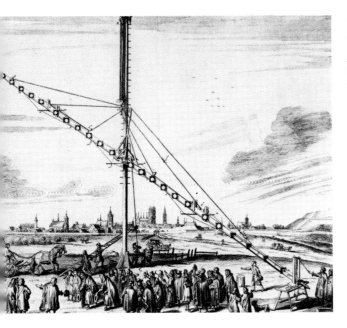

which he then began to use regularly. It occurred to him that the pendulum might be applied to a clock, but he could not persuade a craftsman that this ridiculous idea would work.

The same idea had struck the great Dutch astronomer Christiaan Huygens, the remarkable mathematician and thinker who discovered Saturn's rings, formulated the wave theory of light, improved the telescope and made the two most fundamental inventions in horology. Huygens developed Galileo's mathematical theory of the pendulum and invented the practical pendulum clock. He was granted a patent for the clock in 1657, and the first version was a small wall clock constructed by Saloman Coster, a clockmaker of The Hague, in 1657. The clock, which included other vital innovations that have

Left : An enormously long telescope shown in Helvelius's Machina Coelestis *of 1673.*

theory. Galileo responded with a guarded account of his own views. Kepler replied asking Galileo to carry out some observations for him, and Galileo was so offended that he did not reply for 12 years.

Galileo's main contribution to science was to put modern dynamics in the place of the old Greek notion of conflict between earth and air, but he was much concerned with astronomy. He made a telescope, based on a report he read of a Dutch invention, which he showed to the Venetian senate, who were so impressed they doubled his salary and made his professorship permanent. But telescopes were already being sold at the fairs and in the spectacle-makers' shops in other countries, and soon Venetian spectacle-makers were selling them for a tiny fraction of what Galileo got from the senate.

This made Galileo something of a laughing stock. However, with his telescope Galileo discovered four of the satellites of Jupiter, which he hoped would solve the problem of navigation at sea. He also made observations of Venus that were decisive in proving Copernican theory.

Astronomers found that Galileo's freely suspended pendulum, merely a weight on a thread, was a more accurate timekeeper than any clock. The trouble was that someone had to count the swings and to give the bob a little push from time to time.

One of the astronomers impressed by the pendulum was Johannes Hevelius, who lived in Danzig and owned a selection of sundials and water clocks divided into minutes like his mechanical clocks, as well as a telescope 45 metres (150 feet) long. In 1640 he read a book by Galileo describing the cord pendulum,

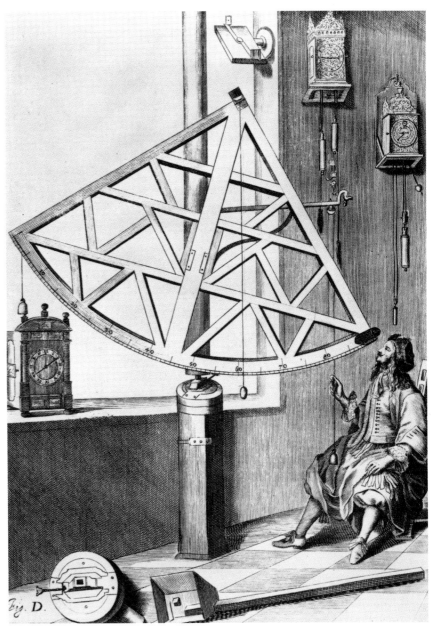

Right : Sketches and notes by Huygens, showing the back of a planetarium clock with a pendulum that has cycloidal cheeks.

Left : A wooden quadrant (c. 1648), illustrated by Helvelius in 1673, with three wall clocks, as timekeeping was vital to accurate observation.

Right : Christiaan Huygens, the celebrated Dutch astronomer, painted by C. Netscher in 1671. He made the greatest single contribution to timekeeping with the pendulum for clocks and spiral balance spring for watches.

influenced the course of horology to present times, is in the National Museum of the History of Science in Leiden, Holland.

In 1673, Huygens published a mathematical treatise on the pendulum called *Horologium Oscillatorium* that caused an extraordinary dispute with one Viviani, friend and pupil of the now dead Galileo. As soon as he heard of Huygens's invention, Viviani claimed that Galileo had invented the pendulum clock, although he had previously written Galileo's biography without mentioning such an invention. One Italian academy even published a picture of the 'clock'. The drawing looks like a garden sundial with a pendulum hanging below it, and it upset Huygens, who protested by letter.

Galileo's 'clock' might have remained a matter of purely academic dispute had it not been for an odd coincidence which occurred about 80 years after Viviani's claim. An Italian professor bought some meat from his butcher and found it wrapped in some pages of an old Latin manuscript. His curiosity aroused, he discovered that they had been written by Galileo. Another visit to the butcher enabled him to recover many more valuable manuscripts and letters which had been hidden by Viviani in a grain bin in his

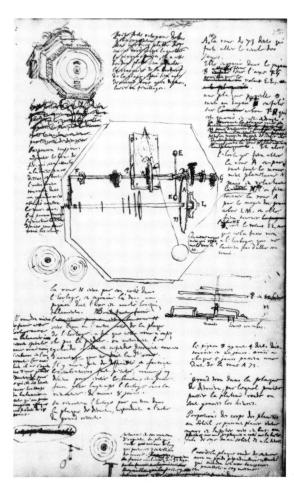

home in Florence and been sold by weight as wrapping paper to the butcher by Viviani's nephew. The nephew had discovered them many years after Viviani's death.

The papers described how Galileo had invented a pendulum clock when he was blind and asked his son Vincenzo to make it. Vincenzo kept putting off the task and did not complete it until some years after his father's death. Then, one day, Vincenzo destroyed most of his clocks in a delirium. The pendulum clock must have escaped, however, as it was referred to in his widow's will.

This rather dubious story was suddenly confirmed in 1856 when a drawing of Galileo's design was discovered. His application of the pendulum to a clock was fundamentally different from that of Huygens, and not as practical, although it contained the germ of the chronometer escapement. Had Galileo not been so secretive, the history of clock design might have taken a different direction.

There is still a joker in the pack, however. Leonardo da Vinci completed several sketchbooks from 1493 to 1494 which show various mechanical devices, including clock escapements, gears to show minutes, springs and fusee, and two drawings of a pendulum attached to a verge escapement. This suggests

that Leonardo may have invented the pendulum clock well before Galileo or Huygens. However, there is no evidence that such a clock was actually made. In fact, Leonardo's 'clock' was probably an automatic device to count the swings of the pendulum, but not relating the swings to time of day.

ISAAC NEWTON

In the latter part of the seventeenth century, the paths of the astrologer-astronomers and mathematicians of the Royal Society converged no longer exclusively on Gresham College, but increasingly on Cambridge University, where Isaac Newton (1642–1727) was from 1669 professor of mathematics.

Newton's *Philosophiae Naturalis Principia Mathematica*, or *Principia*, published 1686–8, transformed the scientific understanding of the universe. *En route* it answered clock-makers' problems concerning the effect of gravity on the pendulum and the cause of the tides by explaining and supplying a formula for the attraction between two masses. Tides were caused by the attraction between the Earth and the moon distorting the oceans. Tidal times were useful information for a clock to give, and the practice became quite common in the eighteenth century because the mechanism needed was almost identical to that for the moon dial.

A pendulum made to a length that beats seconds in London would be faster at the North Pole and slower at the equator owing to flattening of the Earth at the poles. It would be 3 minutes 50 seconds a day slower, so a pendulum could now be used to measure gravity.

For his clocks, Newton is believed to have patronized a clockmaker in Coventry called Samuel Watson (*fl. c.* 1687–1710), who had been appointed Mathematician in Ordinary to King Charles II. He supplied astronomical clocks with an Earth-centred universe to Charles II and to Queen Mary. Watson's clocks were unusual because they showed the full planetary motions. They were probably the first to do so in England since Richard of Wallingford had made his famous astronomical clock about 300 years earlier.

In 1695 Watson made a clock with a heliocentric planetary dial which Newton may have ordered and which in any event he acquired. It is a table or bracket clock with an elaborate dial. The clock shows the day, the month, the principal stars at the zenith (overhead), the position of the sun in the ecliptic, times of rising and setting of the sun, the phase of the moon, the aspects of the sun and moon, and the times of high tide at London Bridge. It also predicts eclipses, and, of course, tells the time of day.

The Watson astronomical clock supplied to Queen Mary has a curious history. It was ordered by Charles II but took six years to make, by which time Charles was dead, James II had abdicated, and William and Mary were on the throne. Watson despaired of selling it and planned to have a lottery, selling 100 tickets at £10 each. Ten numbers would be drawn, each to receive a watch, and from them one would be drawn to win the clock. He also advertised in the *London Gazette* in 1690 his 'curious piece of Clockwork, representing the Motions of the 7 Planets, with their Equations, Aspects, and other Phenomena; and also the Motions of the most remarkable Fixed Stars in the Zodiac, as they shall appear on the 30th of March, 1691'.

There was no need for the lottery in the end, however, as the Queen did eventually buy the clock, for the then considerable sum of £1000. It is now in the Library at Windsor Castle.

As Newton's *Principia* was incomprehensible to most people, models of a heliocentric universe were needed. The first known was made by Thomas Tompion and George Graham in about 1710, and is now in the Museum of the History of Science in Oxford. It is in an octagonal case with a clock dial on one side marked with two rings of figures to show the difference between solar and sidereal time. A handle in the centre of the dial turns the mechanism on the top,

Left : The original orrery or planetarium was made by John Rowley in about 1705 for his patron, the Earl of Orrery, after a design by George Graham.

Right : An orrery clock by the French maker, Raingo, who made a number of them. The orrery on top is wound separately and can be set off by the clock or operated by hand for demonstration.

Left : A Swiss orrery
clock by J. F.
Ducommun of La
Chaux de Fonds. The
Earth revolves around
the sun and the moon
around the Earth.

Right : A longcase
clock by Edward
Cockey of Warminster
with an astronomical
dial, c. 1760, it stands
over 3m (10 ft) high.
It has a year calendar
and shows the sun's
place in the ecliptic,
age and phase of the
moon, equation of time,
and Church of England
Sundays.

which comprises a ball in the centre for the sun, and on the outside other balls represent the Earth and its moon.

Similar instruments became popular by the middle of the eighteenth century, not only as teaching aids but as novelties for the drawing-room. They were known as orreries, after one made in 1716 by John Rowley for Charles Boyle, Earl of Orrery, but are more correctly called planetaria. The orrery or planetarium replaced the armillary sphere, and, as it was expensive because of the gearing, attempts were made to simplify it, notably by the London instrument-maker Richard Glynn, and the clockmaker James Ferguson.

Orreries became more and more elaborate. One made by George Adams in 1740–50 at his workshop, which he indicated by the Sign of Tycho Brahe's Head in Fleet Street, still

exists. It is about 1 metre (3¼ feet) in diameter, and includes Saturn with the ring and five satellites, which completes its orbit in 29 years and 168 days. The gearing has to accommodate this, as well as Jupiter with four satellites, Mars, Earth with its moon, Venus and Mercury, all with widely differing times of orbit.

The artistic culmination of the drawing-room orrery was probably reached in those of the French clockmaker Raingo of Paris, in the early nineteenth century. His style was to mount the orrery on a circular platform of four classical pillars. Between the pillars is a pendulum clock. The planetarium part shows the Earth revolving on its axis round the sun, while the moon turns correctly round the Earth. This part is wound every four years, separately from the clock, but it is worked by the clock, or for demonstration it can be separately operated by a crank. Ten such clocks have been located.

ACCURATE TIMEKEEPING

When the longcase clock became common, clock-owners had to set their clocks by the sundial. Those with more accurate clocks soon began to notice that clock time kept varying from sundial time, even by comparison with the most accurate sundial with its gnomon at the correct latitude. Clock and sundial time coincided at four times, in the middle of April and June, at the end of August and at around Christmas. For the rest of the year the sundial was either fast or slow by the clock. For example, in the middle of February, the sundial was about 14 minutes slow by the clock, and at the beginning of November, about 16 minutes fast.

As it was impractical to alter the clock throughout the year, the clock's equal hours, or mean solar time, gradually became the accepted system in place of the sundial's solar time. In the later sixteenth and the seventeenth centuries it was common to supply a sundial with higher-quality clocks. Also provided was a table, known as an equation of time table, showing the differences between solar and mean solar time on different days of the year. From the late seventeenth century, clocks, even watches, were sometimes made with an equation dial, showing automatically what correction should be applied on the current date.

The accurate clock pinpointed the shocking fact that the sun was not a very good timekeeper. The slightly elliptical and apparently tilted path of the sun made the solar days vary in length, while the clock behaved as if

the path of the sun were flat and circular, to give equal hours. Then the free pendulum showed that the Earth did not spin evenly on its own axis, because the axis itself was always changing.

Astronomers had for centuries been aware of the fact that the most accurate measure of the day was to time the rotation of the Earth, but not from the sun. They identified certain stars that appeared to remain in the same place, and called them clock stars. If a telescope is fixed on, say, the Greenwich meridian, it will sweep across the sky as the Earth turns. If the time is noted at the instant when the clock star crosses a vertical line across the eyepiece, that will mark the beginning of a period of 24 sidereal hours, which will end when the star appears to sweep across the line the next night. The period between one such transit and the next is called a sidereal day.

In order to make transit observations, a telescope was needed. A transit instrument is a telescope that is rigid from east to west but can be swung north to south (up and down the meridian), because a clock star can appear at different heights in the sky. The modern version, the photographic zenith tube, photographs stars in a way that eliminates human error and automatically compares the reading with that of the observatory clock.

As the Earth loses a day every year by going round the sun as well as turning on its own axis, the sidereal year is a day shorter and the sidereal day is almost 3 minutes 56 seconds shorter. Sidereal time is of no value for ordinary everyday use because it takes no account of when it is light and dark. Astronomers have some clocks running at sidereal time and have to correct this to transfer it to mean solar time clocks in order to provide a time service.

The sidereal year used to be measured from the vernal equinox, which slowly changes, so in 1939 it was agreed by the International Weights and Measures Committee to adopt the sidereal year at 1900 as the basic year, giving that the name of the tropical year. A second was defined as a certain fraction of the tropical year.

The invention of the atomic clock had a profound effect not only on the accuracy of timekeeping and astronomical observation, but on the very definition of time itself.

Quartz clocks had already shown that the length of the day changed throughout the year, even after the previously known variations had been allowed for. The atomic clock showed just what a bad timekeeper the Earth was. It gained time in the summer and lost

Right : The transit instrument installed by Halley and used for measuring time by the Royal Observatory from 1721 to 1815. It is mounted to swing only north-south.

Left : A fine equation clock by Daniel Quare, London (c. 1710), which goes for a year on winding. The equation dial is unusual in being separate from the main dial and in the trunk of the case.

time in the winter, the biggest difference in the two rates being about four hundredths of a second. Each year the changes were different, although their general pattern was the same. Reasons suggested for irregularity of the Earth's rotation were the melting of snow every year at the ice-caps, the effects of tides, even the sap rising or falling in trees. The most likely reason is changes in the core of the Earth.

The atomic clock also indicated that the Earth was gradually slowing down. It loses about one year in 5000 million. This is allowed for in calculating time by the world's observatories.

On 1 January 1958, atomic time officially took over from astronomical time. The change was more dramatic even than that. Astronomers abandoned the system of reckoning years from the birth of Christ. They started the new atomic year of time at 0 hours, 0 minutes and 0 seconds on 1 January 1958. They also defined the atomic second.

The Earth could not compete with the atomic clock, and because of its bad time-keeping, corrections had to be made continuously to atomic clocks to keep them in step with the Earth. The system was not satisfactory, as the time service was not able to

Left : Sidereal and mean time dials designed by Joseph Vines in 1836 and made by Walsh of Newbury. There is only one escapement beating sidereal seconds. The small seconds dial on the mean time clock revolves slowly backwards so that the hand can show mean time seconds and sidereal seconds.

Right : A ten-inch (25cm) zenith tube, the instrument that replaced the transit instrument by making automatic records of the transit of 'clock stars' to determine astronomical time at an observatory.

offer long, unchanging periods of atomic time, On 1 January 1972, co-ordinated universal time (UTC) replaced GMT as the basis of 'ordinary' time, and was set precisely ten seconds behind international atomic time (TAI). Ten countries, including the United Kingdom, the United States and Switzerland, adopted the system of saving up the losses and applying leap seconds when necessary. The system is similar to that of adding leap days to make up for the odd quarter of a day of every year. Some countries, including France and Germany, continued with the old system of continuous adjustment, but France came into line in 1978. The first leap second was applied in October 1972, and leap seconds have been applied three times since.

When such accurate time signals are transmitted to different places, the only correction that remains necessary is to deduct the time of transmission at the speed of light or of radio waves (which are the same). Signals are transmitted in this way between the United Kingdom and the United States by bouncing them off Telstar I. In this case all that is needed is a correction of a millionth of a second.

There are plans to mount an atomic clock in a satellite to provide a synchronized time

Left : The original atomic clock which used the emission of caesium atoms to establish a time standard. It was developed by Louis Essen and J. V. L. Parry of the National Physical Laboratory, Middlesex, England, and went into service in 1955.

service over a large area of Earth. Already there is a Japanese wrist-watch on sale that is powered by light. If it could receive a synchronizing signal from such a satellite, it would become the perfect personal timekeeper that never needed winding or a new battery and was as accurate as an atomic clock.

Space travel is utterly dependent upon accurate timekeeping. The visual and audible signals used by control at Houston are familiar to anyone who has seen a space shot on television. When signals are transmitted between Earth and a spaceship hundreds of thousands or many millions of miles away, there is an increasing and appreciable delay. It was noticeable on the voice messages between control and module during the moon missions. Jodrell Bank radio telescope had to wait for nearly two minutes for messages from Pioneer V, over 20 million miles away. When sending signals that guide unmanned space probes to moving targets many millions of miles away through gravitation fields that change their courses, these time lags have to be calculated with great accuracy.

THE INFLUENCE OF THE SCIENTISTS

At one point, scientists and technicians, taking over where clockmakers had left off, seemed to have achieved a miraculous triumph in measuring absolute time. Newton's fundamental clock was the Earth rotating in absolute space. His belief had much experimental support. In the middle of the nineteenth century, a Frenchman, Foucault, hung a pendulum bob from a long cord. He found that, when looking down the cord, the plane of vibration slowly turned like a clock hand. At the North Pole, the pendulum turned round once in a day, like a 24-hour clock hand. This suggested that the pendulum was swinging in absolute space and the Earth was turning under it. At the equator it did not turn at all.

Yet the extreme accuracy of the atomic clock was eventually to prove that there was no such thing as Newton's absolute time. The truth lay in Albert Einstein's theory that beams of light were our ultimate clocks.

While working as a clerk in the Swiss Patent Office at the beginning of this century, Albert Einstein (1879–1955) started questioning Newton's fundamental beliefs by putting himself in the place of an observer of events. If you could look at the dial of a clock that was rushing away from you at an enormous speed, at each jump of the seconds hand the

Left: The quartz module has been miniaturized to watch size and can be powered by a solar cell to give indefinite operation. This version is the Synchronar.

Below: A Patek-Philippe quartz crystal clock operated by a solar cell that turns automatically to seek the best light source.

clock would be further away and the light by which you saw it would take longer to reach you. The clock would appear to go slower because there were larger intervals between the jumps of the seconds hand. To put it another way, if you left the clock showing 02.00 and rode away on a light beam, the time by that clock would be frozen and remain at 02.00 for as long as you travelled at the speed of light. There would be no need to leave the clock behind; you could circle around it.

Einstein eliminated the idea of instantaneous influences. There is no force that can act instantaneously anywhere. It cannot act at a speed greater than the speed of light, which is the absolute constant for everyone.

If there were three men each with a clock on three stars, number one might see the clock of number two as losing an hour a day and that of number three as gaining half an hour a day. Number two might see the others' clocks as losing half an hour and losing a quarter, while the third man's observations would be different again. No one has the absolute time. Each observer carries his own time around with him when travelling at speeds approaching that of light, 186,000 miles a second. There is no universal point of time called 'now'. Just as things do not exist un-

less they have some small span of time in which to exist, so time itself cannot exist unless it has some matter buried in space in which to exist. Thus every world has its own 'here and now'.

The idea of time that can go at different rates and varies from place to place is still difficult to grasp. Even Newton surely would have gasped at the notion that time is effected by gravitation.

Newton's proposition that weight could change and would be different in different places was just as shocking at the time as Einstein's similar views about time. Yet today we are used to astronauts floating weightlessly in space. At some future date the fact that time can speed up or slow down may be borne upon us dramatically. To physicists dealing with sub-atomic high-velocity particles, it is already a familiar idea. For example, a positron is described as an ordinary electron 'travelling backwards in time'.

Einstein's theories of relativity have been tested over the years as the means became available. The first successful test concerned an explanation of the erratic behaviour of Mercury, which had been bothering astronomers for centuries. Another was conducted at the Atomic Research Establishment at

Right : The Triumph of Death *by Pieter Breughel suggests the inevitable progression and absolute quality of time according to Newton.*

Left : The painting Disintegration of the Persistence of Memory *by Salvador Dali suggests the relativity and impermanence of time according to Einstein's theory.*

Harwell in 1960. An atomic marker is a radioactive substance that can act like a clock or counter. A piece of such a substance was embedded near the axis of a flywheel and another piece near the rim. The flywheel was spun so that its rim was travelling at about 500 miles an hour. The 'ticks' of the two 'atomic clocks' could be picked up by counters and compared.

The difficulty was in making the comparison. It was done by what is known as the Mossbauer effect, which is comparable to the method of the piano tuner when he listens to the beat between two almost identical notes. At Harwell, scientists were able to measure the differences between atomic vibrations of 1,000,000,000,000 and 1,000,000,000,001. They found that the 'clock' at the rim was losing three seconds in a million years in relation to that at the centre because of their relative speeds. Not much, but it supported the theory.

One objection to Einstein's theory is known as the clock paradox. Passing of time leaves a permanent effect. If you went on an excessively fast space trip for a long time, when you returned to Earth, you would expect the clocks on Earth to be slow compared with yours, and people on Earth would expect yours to be slow compared with theirs, which is contradictory. The paradox is countered by calculations involving acceleration.

Einstein had a profound effect on modern thinking. Not all physicists have accepted his theories, however. One prominent dissenter today is Louis Essen, the man most responsible for the atomic clock. Since the 1950s, Dr Essen has contested the basis of the special theory of relativity, mainly because its thought-experiments are not self-consistent. It could be that the latest universal theories, proposing an all-pervading high-frequency electro-magnetic field, will bring Newton's absolute time into prominence again.

THE INFLUENCE OF CLOCKMAKERS

The first industrial revolution involved the conversion of fossil fuels into power to replace wind and falling water and the muscles of horses and men. The second industrial revolution, now in progress, is in the harnessing of electronics to supplement or replace man's brain-power and skills. The reader will know that the earliest automata were wooden models of men which struck bells and replaced real men. Automation expressed in another way is automatic control. The first sophisticated application of automation in

this sense was by the famous watchmaker A.-L. Breguet.

Between 1809 and 1836, the firm of Breguet made seven special clocks, each of which was accompanied by a watch. The owner put his watch in a slot in the top of the clock at night and the clock automatically set the watch to time and, in later versions, wound it, calculated how much it had lost or gained and altered the regulator accordingly to adjust it.

Although with these clocks, which he called *pendules sympathiques*, Breguet intended to demonstrate his great superiority in horology, in fact they realized the primary principle of automation, to take a sample of an error and feed it back to control, which would then make a suitable adjustment. He used what is now called negative feedback.

At all ages and at all times, there is someone trying to think of a new way of measuring time, which may change history or may become no more than a novelty. The clockmaker's influence is everywhere. The universal joint, the cam, the cardan shaft and the differential gearing now used in automobiles, spring-power, the thermostat, various types of instrument dial, the endless chain, the governor, the original memory device, now used electronically in computers, the seeds of automation, all these advances were the ideas of clockmakers.

The first domestic article successfully mass produced was the grandfather clock in America. Thomas Tompion was previously an early pioneer of batch production in England. Henry Ford started as a watchmaker, and so did William Morris, both pioneers of the motor car and of mass production.

For us all the progress of time is inexorable. Man is a being who can only grasp events in a time order, and it is given only to a few to experience the merest glimpse of the great underlying reality of space time. We are stuck in this moment of time, which becomes the past even as we think of it, like a flash of lightning that is a past event before it is seen.

If we try to stop, to cling to this moment and resist change in a literal sense, we die. If we do it in a metaphorical sense, we get out of touch with reality. There is no alternative but to travel with the clock's 'now'. For us it is an absolute. As A. W. Watts wrote in *The Wisdom of Insecurity*,

You cannot compare this present experience with a past experience. You can only compare it with a memory of the past, *which is part of the present experience.*

Left : Albert Einstein (1879–1955), a picture taken in 1932.

XI

GREAT CLOCKS OF THE WORLD

STONEHENGE

The majestic ring of stones at Stonehenge on Salisbury Plain in Wiltshire, England, has been associated in legend with the magician Merlin and with the Druids, but is much more likely to have been a Stone Age equivalent of a modern astronomical observatory and time service. It may have been a Druid temple, but not until a long time after it was built; any association with Merlin is more tenuous. Three successive Stonehenges have been identified on the same site and carbon dating has put the age of the oldest at about 1900 BC, in the late Stone Age. It was then a 350-metres (1150-foot) circle of 56 holes (now called the Aubrey circle after its discoverer), surrounded by a ditch and banks outside and inside the ditch. Two stones were set up as an entrance and a third, the heel stone, set about 90 metres (295 feet) outside the circle.

Around 1750 BC, the Beaker people put up two inner rings of blue stones, also with an entrance towards midsummer sunrise. Sometime after 1700 BC, the blue stone circles were removed and a ring of 30 huge Sarcen (natural sandstone) stones erected with lintels or cross-pieces. Inside were five even bigger Sarcen trilithons, each being two huge uprights with a lintel held on top by a form of mortice and tenon, the five being in the shape of a horseshoe. The space between each pair of uprights is only about 25 centimetres (10 inches). The association of the heel stone with sunrise is well known. Other alignments had been noted by the archaeologists R. S. Newall and C. A. Newham; moreover, Mr Newham had come across a reference by Diodorus the Greek in his *History of the Ancient World* (*c.* 50 BC) to a temple, spherical in shape, in the land of the Celts, which Apollo (the sun god) visited every 19 years to play the cithara and dance through the night from the vernal equinox (first day of spring) to the rising of the Pleiades.

Stonehenge is orientated so that the setting of the sun at the winter solstice (shortest day) is seen through the great central trilithon from the centre of the ring, and, in the opposite direction, the sun rises over the heel stone at the summer solstice (longest day). The activities of present-day Druids have publicized the dramatic picture of the flash of the midsummer sun as it appears. A 1.8-metre (6-foot) man today sees the tip of the heel stone on the horizon. Gerald S. Hawkins, professor of astronomy at Boston University, has calculated that the flash would have appeared three-quarters of a degree to the north in 1800 BC and that if the heel stone, which leans today, had been upright, the disc of the sun would have stood exactly on the tip of the heel stone. It was a remarkable feat of engineering, to raise a 35.7 tonne stone and to sink it in the ground so that the tip was within a few centimetres of an astronomical alignment. Comparable accuracy was needed for the erection and alignment of the other, some even larger, stones. By means of a brilliantly conceived computer analysis, Professor Hawkins discovered that the first Stonehenge had 11 viewing points, each of which was paired with another to provide 16 alignments indicating 10 extreme positions of the sun and moon. The third Stonehenge's trilithons and Sarcen ring gave four more sun and four more moon pointers. Even today, standing in the middle of the mostly fallen ring, one is forced to look through paired

Left: One of the dials of the Rouen clock in the rue Grosse Horloge. The dial on the other side is identical. The old wrought iron movement of 1389 can still be inspected by visitors.

STONEHENGE

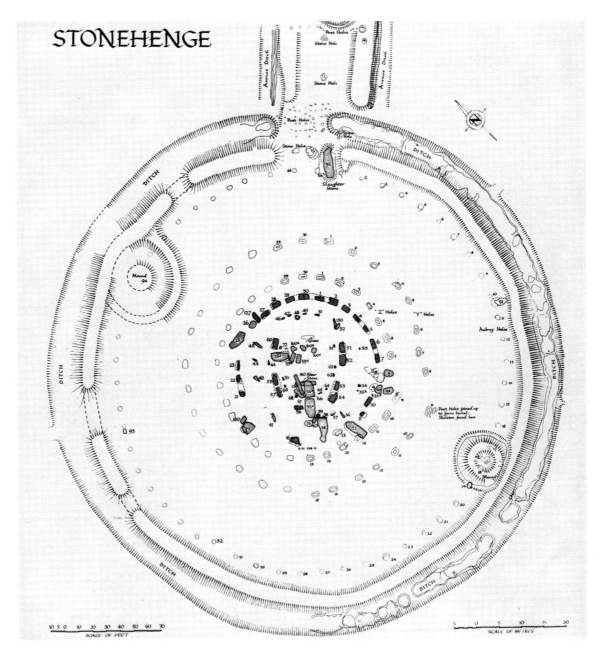

SCALE OF FEET

SCALE OF METRES

archways that have significant astronomical alignments. The odds against the stones of the two Stonehenges having been erected at random are ten million to one. The eight key positions of the sun are its rising and setting at the two solstices (midsummer and midwinter) and the two equinoxes (the midpositions when spring and autumn start). The sun's movement is consistent; it rises and sets to a northern limit in the summer and a southern limit in the winter. The moon's is so complicated that even astronomers have difficulty in visualizing it. Contrary to the sun, it goes south in the summer and north in the winter, but it moves between two extremes in each direction, instead of to a fixed limit like the sun, over a period of 18.61 years. It is like a pendulum gradually reducing its swing and increasing it again over this period.

There are two, now filled-in, holes, one each side of the heel stone, that once held stones showing the limits of travel of the rising winter moon over the period. Limits of the summer moon are indicated by the sides and diagonal and corresponding Aubrey holes. The inner raised bank around the first Stonehenge acted as an artificial horizon, because variations in the natural horizon affected observations. The mystery of the 56 Aubrey holes may also have been solved by Professor Hawkins in a later investigation. He computed the positions of the moon from 2001 to 1000 BC (using 40 seconds of computer time). The midwinter moon was found to rise over the heel stone 32 times at intervals of 19 years and 20 times at intervals of 18 years. Other moon alignments follow this cycle, including eclipses of the sun and moon – potent

Left : A plan of Stonehenge, showing the outer bank, probably used as an artificial horizon, and the Aubrey holes just inside it.

events at the time – so moonrise over the heel stone was a signal that an eclipse was possible (about half of all eclipses were above the horizon and could be seen). If a cycle of 19 years had been taken for predicting when such a moonrise would occur, a correction would have been necessary every 56 years. This was the clue to the purpose of the 56 Aubrey holes. One way in which they could have been used was to place six stones in holes at intervals of 9, 9, 10, 9, 9 and 10 holes, alternate stones being different – say one being white and the next black. A white stone, say, would have been placed in the hole opposite the heel stone at the year of an eclipse. If all the stones were moved round one hole clockwise every year at the summer or winter solstice, the white ones would appear opposite the heel stone at intervals of 19, 19 and 18

years, which is equivalent to a cycle of 18.67 years, as compared with the true one, which is today known to be 18.61. Any stone opposite the heel stone would indicate that the full moon would rise over the heel stone, and other predictions have also been found possible with the holes. Checked against actual eclipses from 1610 to 1425 BC derived by a modern electronic computer, the Stone Age computer was right 14 times out of 18. The original discovery of the lunar cycle is attributed to the Greek astronomer Meton, who noted that every 19 years full moons occurred on the same days of the month. Later the years were given numbers from 1 to 19, called Golden Numbers, to enable the dates for Easter to be calculated. Some early mechanical clocks indicated the Golden Number.

Right : Only the remaining Sarcen trilithon stones and the remains of the bank and ditch are still easily visible. The Aubrey holes are filled in.

THE STRASBOURG CLOCKS

The first clock built in the Gothic cathedral of Strasbourg, now an industrial city near the Rhine in eastern France, was one of the wonders of the Middle Ages. It was built about 1354 and was an immense structure about 18 metres (59 feet) high and 7.5 metres (25 feet) wide, to judge from the lead anchorage plugs that still remain in the walls. All that now remains is a mechanical cockerel that stood on the top and crowed at noon, opening its beak, putting out its tongue, spreading the wrought-iron feathers of its wings, and flapping its wings. When the cathedral's second monumental clock was completed in 1574, the cock was still working so well after 220 years that it was incorporated. The original cock was sent to the City Museum, with most of the old movement and the astronomical dial.

Who built the first clock and made the cock is unknown. Alfred Ungerer, a clockmaker whose family is related to the maker of the present Strasbourg clock (and whose company still looks after it), has suggested that the maker of the first clock might have been Johann von Hagenau, who in 1383 built the astronomical clock in Frankfurt Cathedral. The complex gearing of the second clock of 1574 was calculated by Conrad Hazenfratz (1532–1601), professor of mathematics at the university. The makers were Isaac I (1544–1620) and Josias (1552–c. 1575) Habrecht, two celebrated members of a family of clockmakers from Schaffhausen, Switzerland, whose father, Joachim, had built the tower clock in Solothurn in 1545. There is an engraving by Stimmer showing how the second astronomical clock appeared, with the cock on the top of a tower by the side of the central pillar. The clock's fame was so great that Isaac was commissioned to build two working models, one for Pope Sixtus V and the other for the King of Denmark. The first of these is in the British Museum in London, and the other in Rosenborg Castle in Copenhagen, Denmark.

The gilded copper case of the British Museum model is 1.5 metres (5 feet) tall. There are three dials, one above the other, with two automata above the lowest, in the form of two cherubs; on the hour one of these moves his scythe and the other turns a sandglass. Above the dials are four galleries of automata. On the lowest, symbolic days of the week pass in front of a star-spangled blue sky. On the next, angels pass at each hour in front of the Virgin and Child. The next tier holds the quarter bells, which are struck by figures representing the four ages of man, the child, youth, middle-aged man and ancient. At the very top is a cock that crows at the hour, but it is not as elaborate as the original one and only flaps its wings.

The lowest dial is the largest and has as its outside ring an annual calendar engraved with the months and days, which revolves once a year and indicates the date and also the Dominical Letter against a small pointer. In the centre is a dial showing 1 to $29\frac{1}{2}$ days and a revolving moon disc that shows the moon's age and also its phase. Between the inner and outer dials, the signs of the zodiac are engraved. An extension of the moon phase hand shows the position of the moon in the zodiac, and a sun hand, turning once a year, shows the position of the sun in the zodiac.

The next dial up shows quarter-hours, with divisions every five minutes, a not uncommon feature of the more advanced early clocks before the concentric minute hand became possible. The top dial indicates hours on a double-XII dial, with 24 hours also shown in Arabic numerals. The hand has a sunburst at one end, which indicates time of day. The long tail carries an effigy of the moon and points to the astrological hour,

Far left : The sun setting in winter at Stonehenge. This could be used to determine the winter solstice, the shortest day, and the equinox, when the lengths of daylight and darkness were equal.

Left : The famous astronomical clock in Strasbourg Cathedral was first constructed in 1354. The present clock was completed in 1842.

which was 12 hours out of phase with the time of day, starting at midday. It indicates the same number as the sun on the double-XII dial, of course.

The other clock, in Denmark, was made later and is different in several ways, for example in having a celestial globe in front of the plinth on which it stands, although it is possible that this was added later. The position of the sun in the zodiac is shown by the globe instead of on a dial. The main dial gives the date of Easter, but not automatically, as well as various church feasts, the Golden Number, the Dominical Letter, and the years of leap years. It bears an inscription that translates as 'Isaac Habrecht, Clockmaker and citizen of Strasbourg, 1594'. Habrecht only followed the design of the cathedral clock in broad outline. The Rosenborg model has three upper dials, the largest one being the earliest known with concentric hands, one showing the hours and the other the quarter-hours (in fact, it could show minutes, but they are not marked). Two smaller dials at the top indicate the days of the week and the four ages of man; figures representing the four ages of man strike the quarter bells.

Whatever the second Strasbourg Cathedral clock itself was like, it must have been a truly amazing construction by any standards. In the later eighteenth century it stopped, and no one could be found with enough knowledge to put it right. The silent giant clock was visited, so the story goes, by a small boy named Jean Baptiste Schwilgué (1776–1856), who was so fascinated by it that he assured the startled guide that when he grew up, he would set it right. He did not, and the old movement was removed to the Strasbourg Horological Museum, but he did devote most of his life to designing and building an even more complex clock.

He had set up in business with a partner, Rolle, making turret clocks and portable weighing scales which he designed. During this period, he began to study the mechanisms of the second Strasbourg clock and, in 1821, he devised gearing for a calendar that would show Easter and the movable feasts of the Church automatically, something that had previously been done, as far as is known, only by Giovanni da Dondi. It is unlikely that Schwilgué knew anything of Dondi's achievement. In any case, in Dondi's time, the Julian calendar provided for a leap year every four years, but after 1582, when the Gregorian calendar was introduced (not until 1752 in England and the English colonies of America) the leap day was omitted every

*Left : An engraving of
1608 showing Isaac
Habrecht, one of the
builders of the second
Strasbourg clock.*

*Left : The amazing
cock automaton, which
was on top of the first
Strasbourg clock, still
exists.*

*Right : An engraving of
the last Strasbourg
clock after its
rebuilding in 1842,
drawn by Wissant and
engraved by Rouargue.*

century except every fourth century, which made Schwilgué's mechanism much more difficult to devise.

Schwilgué's amazing calendar mechanism was so admired that the Mayor of Strasbourg was encouraged to ask him to quote for the repair of the 1574 cathedral clock. Schwilgué made three proposals: to rebuild the old clock, to modify it, or to build a new one. The last proposal was accepted, but shelved for financial reasons until 1838. After work was started, completion was timed to coincide with the centenary of the Scientific Congress in France in 1842.

Left: The carillon clock, based on a Strasbourg design, was made by Isaac Habrecht for Pope Sixtus V, and is now in the British Museum, London. It stands 1.6m (5 ft 2 ins) high.

Right: One of Dondi's sketches of his clock, this was the first to show an escapement. It has a crown shaped circular balance, and 'pudding basin' weights.

Schwilgué's clock is the one that is admired by thousands of visitors to Strasbourg Cathedral now. In front of the lowest dial is a globe that has a dial on its north pole (facing to the front) indicating sidereal time by rotation of the globe, which is corrected for precession of the equinoxes (movement of the poles in a small circle over 26,000 years, involving gearing of 9,451,512 to 1). The dial behind it gives time of day, times of rising and setting of the sun and the position of the moon. The outside ring is a perpetual calendar revolving once in a year, the date being read from an arrow held by a figure on the left. On 31 December every year, the ring is adjusted automatically so that Easter and other movable church feasts are shown opposite the days on which they occur. A panel on the left of the main dial contains gearing for calculating the date of Easter from the solar cycle of 18 years (Dominical Letters indicate the dates on which Sundays fall during the cycle) and the lunar cycle of 19 years (Golden Numbers indicate when the full moon falls over the cycle). This information is transmitted to the main dial. On the right is another panel, with mechanisms that calculate the effects on the measurement of time of the inclination of the Earth's axis, the elliptical path of its orbit, and the angle between the sun's and moon's orbits, and transmits this information to the planetary dial, with signs of the zodiac around it, near the top of the clock. The planetary dial shows the movements of Mercury, Venus, Mars, Jupiter and Saturn with much greater accuracy than had been done before. Between the two panels with their calculators is the mean solar clock dial, giving time of day. The lowest automata are below this, comprising a procession of the appropriate deities for each day of the week. On each side is a small boy, one striking a bell and the other turning over a sandglass at the quarterhours.

The lower of two galleries at the top shows Death in the middle as the hour, while the four ages of man pass him representing the quarters, the child striking one note as the first quarter, and an old man on crutches striking four before the hour. At the hour, a procession of the 12 apostles passes in front of Christ in the upper gallery, while he raises and lowers his hand to bless them. At the very top is a cock, which has all the movements of the original one and crows at noon. The cock is supposed by some to relate to St Peter, who heard it crow thrice, but it may have been incorporated originally simply as a symbol of a new day.

DONDI'S CLOCK

The earliest detailed description of a clock with working drawings is in a manuscript written by Giovanni da Dondi in about 1364. It describes a complex planetary clock that was constructed between the years 1348 and 1364 and was 'so great a marvel that the famous astronomers of distant countries came with all reverence to visit Giovanni da Dondi and his work'. The clock was made for the library at Padua, in Italy. It remained there until 1529, by which time it had become neglected and out of order. The Holy Roman Emperor Charles V had it put right after considerable difficulty in finding a craftsman to do so, then took it to Spain. It is believed to have been destroyed there during the Peninsular Wars.

Dondi held the professorships of astronomy, logic and medicine at Padua, and of medicine at Florence, at a time when it was possible to have gained all the knowledge that was available in the Western world.

G. H. Baillie was the first horologist, in the 1930s, to study what may be Dondi's original manuscript, in St Mark's Library in Venice. Five copies are at present known, three in Italy and two in England. H. Alan Lloyd later studied the manuscript and initiated the making of a remarkable reconstruction in 1961 by Peter Hayward, of Thwaites and Reed Limited, the oldest-established English clockmakers, following the original instructions. It was thought to be the first full astronomical clock to be made in Britain for 250 years, and went to the Smithsonian Institution, Washington, D.C., where it is on display. Other reproductions have been made since.

The clock was seven-sided, about 1 metre (3¼ feet) high, and astonishingly advanced for its time. As well as time of day, including minutes, it showed the motions, as understood at the time, of all the known planets (which then included the sun and moon), and provided an automatic calendar of the movable as well as the fixed dates and feasts. A clock of such complexity of gearing was not repeated until two centuries later, by Baldewin of Kassel. In recent times, the only comparable clock is that of the Dane Jens Olsen. Dondi did not trouble to describe how the ordinary clockwork was mounted in

Below : One of the reconstructions of Dondi's clock, in this case made for the Science Museum, London.

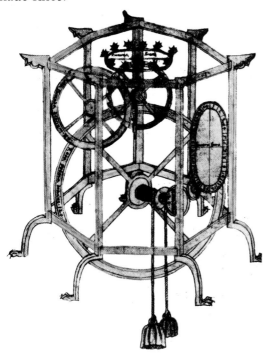

the frame, although he gave details of the toothed wheels and included a verge escapement with foliot balance with a two-second beat, with the earliest known drawing of a mechanical clock escapement. This operated a revolving 24-hour dial showing time against a fixed pointer, which Dondi stated was the 'common clock' of the time. Italian hours were then calculated from 1 to 24, starting at sunset, so a clock had to be reset daily. Dondi commented that it would be more convenient to set his clock by the astronomical method, starting each day at midday. At the top of each of the seven sides was a planetary dial of about 30 centimetres (1 foot) square. Below one of these was the 24-hour time of day dial, revolving anti-clockwise, which had wings at the sides indicating times of the rising and setting of the sun and tables divided into the months and days of the Julian calendar. For the fixed calendar and saints' days, Dondi used a broad horizontal band wheel inside the main frame below the dials, which had 365 teeth in its upper edge and was turned once in a year by the 24-hour dial. Lengths of days and Dominical Letters (used for fixing Easter Sundays), saints' names and dates were engraved around the band, the current ones being shown through a slot in a panel.

A calculator for five movable feasts was operated by gears and chains in a three-compartment rectangular box. This worked out the date for Easter, on which the other movable feasts depend, using the Golden Number (a cycle of 19 years when the full moons are repeated), the Dominical Letter (a cycle of 28 years when the days of the month are repeated), and the Indication (a Roman period of 15 years). In Dondi's day, the planets were assumed to revolve around the Earth in a circle, and at the same time to revolve in smaller circles called epicycles, so that the centre of each epicycle rotated round the Earth. Also, to try to make the theory meet the facts, the Earth was not at the centre of the orbits of the epicycles.

The main dial shows the sun, and the others the moon, Mercury, Venus, Mars, Jupiter and Saturn, those of the moon and Mercury having the exceptional sophistication of pairs of oval gear wheels, one fixed and the other revolving around it, to provide the large eccentricity of the orbits. One more dial, in the frame on the opposite side from the 24-hour dial, is what is called 'the dial of the dragon's head and teeth'. It was a legend at the time that a dragon swallowed the sun and moon occasionally, which was the cause of the eclipses. The orbit of the sun is inclined at 5° to the orbit of the moon and the points of intersection are called the nodes. When the sun and the moon and one of the nodes are in alignment, there is an eclipse, and it was found by observation that this occurred about every 19 years. In Dondi's clock, a hand turning every 18 years, 7 months, 14 days, enables eclipses to be forecast. (This is remarkably near the true cycle, now measured as 18.61

244

years.) A number of corrections had to be made to the clock according to Dondi's instructions, the main one being to stop it on leap days, but to omit this every 36 years. Subsidiary corrections had to be made to the planetary dials at intervals, at an interval of 144 years for one of the Mercury corrections, and Dondi recommended that a reliable clockmaker should keep accurate records of such corrections. The clock frame was probably cast from bronze and the other parts made from brass. There were no iron parts. Dondi's instructions explain how to make even the most simple parts, giving the thicknesses of metals, where to drill holes, lengths of studs, and so on, but his dimensions were primitive by modern standards – the thickness of a thumb's breadth, the thickness of two fingers, the size of a goose quill.

To indicate how advanced the clock was for 1364, these are the dates of the next known clocks with such features: setting of the sun, 1450; length of daylight, 1450; continuous recording of minutes, 1550; dial of the nodes, 1550; paths of the five planets, 1561; conversion of mean to sidereal time, 1565; elliptical orbit of the moon, 1779; perpetual calendar for Easter, 1842. The last date is that of the third Strasbourg clock, the first movement of which was completed in 1354, before Dondi's clock, and may have preceded him with an annual calendar, the day of the month, the Dominical Letter, and the fixed church feasts. Dondi used pinions with teeth (or leaves, as they should be called), unlike the makers of the large wrought-iron cathedral clocks of the time, who favoured the more easily made lantern pinions, like squirrel cages. His wheel teeth were triangular, like those of early Islamic instruments. The clock itself was such a primitive and inaccurate mechanism that it may seem strange that it was loaded with so much accurate astronomical work. The reason why it was successful (apart from the corrections that had to be made by hand from time to time) was that every planetary gearing (except that of the moon) was moved one tooth each year, and because the 24-hour dial was set daily, it was reasonably accurate over a year. The only comparable known clock of the time was designed by Richard of Wallingford and installed in St Albans Abbey, in Hertfordshire, England. In the Bodleian Library at Oxford there is an illuminated manuscript (dated *c.* 1350) describing this clock and an equatorium. It was an astronomical clock with oval gears and triangular teeth and may have been similar to Dondi's.

THE LUND CLOCK

The original astronomical clock in the cathedral at Lund, in Sweden, was completed about 1380 and was similar in various ways to the first astronomical clocks of Strasbourg, Danzig and Lübeck, which were of the same period. It stopped working at the beginning of the seventeenth century and was replaced in 1623 by what seems to have been a humble substitute. The original clock had a complex double dial, that at the top being astronomical and that at the bottom a perpetual calendar, with automata above each. What was left of the original clock and dials was dismantled in 1837 and by 1907, when studies were begun with the object of restoration, the only parts that could be recovered where the remains of the hands, three toothed wheels, the frame of the calendar dial, and some small pieces of sculpture.

Using these, and working from valuable drawings discovered in Stockholm's archives, Theodore Wahlin, architect to Lund Cathedral, and Bertram-Larsen, Copenhagen clockmakers specializing in such restoration, built a reproduction of the original astronomical and calendar mechanisms. This reproduction is contained in an oak case about 8.5 metres (28 feet) tall against the foot of the cathedral

Right : The astronomical clock in Lund Cathedral, dated from 1380, was destroyed and was rebuilt early this century. A movement of 1706 operates it.

tower. The dials are actuated by a movement made in 1706 and installed in the tower, where it strikes the hours twice at the hour and sounds the quarters on three large bells.

The calendar dial is moved daily and shows the month, date, and day, with the appropriate corrections for it to be accurate from 1923 to 2023. It also gives the Dominical Letter, the Golden Number, the solar cycle, the Epact, and the Indication, with the dates of movable feasts. In the centre is an oak statue of Saint Laurent, patron saint of the cathedral. The astronomical section has a double-XII dial around which a hand bearing a representation of the sun moves in 24 hours. Another hand, representing the moon, turns once in 24 hours 50 minutes 30 seconds. At its end is a ball, half silvered and half black, which rotates once in 29½ days to show phases of the moon. On this dial is an eccentric ring bearing the 12 signs of the zodiac. Where the outer edge of the ring intersects the sun and moon hands is an indication of the progress of these bodies around the ecliptic. On the dial itself are circles representing the equator and the two tropics, with a darkened area to indicate the limits of dawn and dusk at the latitude of Lund, as well as the hours of rising and setting of the sun and moon. Automata that appear on a balcony between the two dials are heralded by trumpeters to the left and right. A herald appears brandishing a sword and leading a procession of the Wise Men, who bow to the Virgin in the middle of the balcony in passing, and are followed by their train, while a hymn tune is played on bells. Above the astronomical dial, two equestrian knights charge each other at the hour.

Below : The earliest known quarter striking train, dated 1389, which is in the Rouen clock.

THE ROUEN CLOCK

One of the most important large medieval clocks still surviving is in Rouen, France. The Rouen clock is also the earliest extant designed to strike the quarter-hours. According to records in the city archives, in 1389 the burghers approached a man, probably a blacksmith, and offered him the task of building a mechanical clock. Before then, the clocks had been hydraulic. The man turned out to be insufficiently skilled, so 'the work of this *auloge* was taken from Jourdain de Leche and entrusted to Jehan de Félains'. It is recorded that de Félains took so much trouble over the Gros Horloge, as it is called, that he used up all his own money as well as that made available to him for building the clock, ending his life as a pauper. However, he completed the clock in September of the same year, 1389, and when the administrators saw it they awarded him an indemnity of 70 livres. De Félains was made keeper of the clock and looked after it until he died in 1408. His wife was also granted an indemnity for 'turning' the clock two or three times a week in a way that would avoid damage to it. What this meant was not appreciated until the late F. Knowles-Brown suggested that to put the clock right if it were fast would mean turning back the wheels after lifting the verge out of engagement with the crown wheel, a dangerous procedure.

The clock is in the city belfry at the end of the rue Grosse Horloge (in ancient documents the clock is called the *Gros* Horloge, although *horloge* is feminine), adjoining a bridge over the street in which are mounted two dials. The wrought-iron movement is large, the largest in existence, and is about twice the size of that of the 1392 Wells clock, being 2 metres (6½ feet) wide and almost that in height. The great corner posts are angled and made like stone buttresses, with particularly fine Gothic mouldings, although there are signs of the top decorations' having been removed. Originally, the clock struck the hours and quarters and had no dial. Now there is an internal moving dial turning once in two hours and indicating the time against a fixed pointer. Although this bears de Félains's name and the date 1389, it was added later. The original foliot was replaced by a pendulum in 1713. The layout of the clock is shown in the diagram, the top gear train, set at right angles to the others, being the going (timekeeping) train. As in other large clocks of the time, the parts of the frame are riveted or pinned together, there being no screws. The arbors are octagonal in section

Right : The other Rouen clock dial (see page 234). At the top is a half black Moon which rotates to show the phase, and at the bottom the monthly signs of the zodiac. The carved wavy lines around the dial are a common symbol for clouds.

and the spokes of the wheels are lapped over and welded to the inner edges of the rims.

In the first half of the sixteenth century, when the bridge was built over the rue Grosse Horloge, the old movement was taken from the belfry, mounted in the bridge, and provided with two dials, one on each side of the bridge. Around each clock dial are two rings of sculptured wavy lines, an ancient symbol for clouds sometimes used around the Earth in the centre of early astronomical dials to represent the first of the celestial spheres. The Rouen dials are sun-centred, with an image of the sun in the middle and one of the

Earth on the tail of a hand, which suggests that they were once 24-hour dials, although they are now orthodox 12-hour dials. The dials have been very heavily restored over the years. About 1893 the old movement was removed and a small flat-bed movement installed in its place to work the dials. There is a lunar sphere above the dial to indicate the moon's phase, which was added at this time, or perhaps earlier. The 1389 movement was returned from the spacious room above the bridge to its constricted belfry but the old task of striking the hour and quarter bells is now done electrically.

THE SALISBURY CLOCK

The oldest existing clock in England is the medieval iron clock in Salisbury Cathedral, Wiltshire. It is virtually complete and is almost certainly the oldest clock in the world still working, since its excellent restoration in 1955. The Cathedral accounts for 1386 include a document about providing a house for the clock-keeper, so the clock must have been made in that year or earlier. Ralph Erghum was then Bishop of Salisbury. He moved later to Wells Cathedral in Somerset, and a clock that still exists there was in recent times found to be almost certainly made by the same craftsman, most probably Johannes Lietuijt.

The Salisbury clock has no dial, the only medieval clock remaining, at least in Britain, without one, the hours being sounded on a separate bell by the clock. In 1931 the clock was cleaned up and put on exhibition in the north transept, but not working. At some time it had been converted from verge and foliot control to pendulum, but both were missing. In 1956 the clock was completely restored and X-rayed to identify original and converted parts. A verge and foliot escapement and frame parts were hand-wrought as they would have been made originally, and coloured dark brown to distinguish them from the original parts. The clock was set up in the nave with new dressed stone driving weights made in the cathedral workshops. The clock strikes on the Bishop's Bell, which warned the Bishop of approaching services in former times.

The 'birdcage' frame of the clock which is made from wrought iron, is 1.25 metres (4 feet) high, by 1.29 metres (4-1/6 feet) wide and 1.06 metres (3½ feet) deep. Some of the forged bars of which it is constructed are riveted and others are held by slots and wedges, familiar in old furniture. The cage is divided into two parts, one for the time-keeping train of gear wheels and the other for the striking train. The trains are set end-to-end, so that all the arbors pivot into end-frame bars and a central one. The weights drive through ropes coiled around wooden barrels, which are wound by large wheels like steering wheels. The number of hours struck is controlled by a count wheel with appropriate slots on its outside edge. It is internally toothed, being driven by a projecting pinion. So that the spokes of the count wheel do not butt against the pinion, they are cranked, a feature found on other medieval clocks throughout Europe. Striking is slowed by a fan or air brake on the striking side of the

frame. There are punch marks on the wheel rims indicating where the teeth were marked out and cut by hand. The pinions are so-called lantern pinions, like heavy hamster cages.

These early clocks had to be reset to time against a sundial at regular intervals because the verge and foliot was a poor timekeeper. They were probably set to gain, so that the foliot (the horizontal swinging bar with a regulating weight at each end) could be stopped until the hour was struck at the correct time. To adjust a clock to a time ahead was a dangerous procedure. The verge, which was hung from a cord, had to be disengaged by lifting it so that the crown wheel would start to spin and could be braked by hand. To pull up the verge carelessly, or let the crown wheel slip, would cause the driving weight to crash to the ground. Dropping in the verge while the crown wheel was moving would strip the teeth, with the same result. The clock had a plate, with lifting pins to let off the striking, attached to the timekeeping great wheel via a ratchet. There were indications on the plate of there once having been a 24-hour dial. The ratchet connection would then have been necessary to synchronize the dial and striking without disengaging the escapement.

Right: The restored foliot of the Salisbury clock, with its verge and crown wheel, the lantern pinion of which is driven from the great wheel, attached to the barrel. Dated 1386, it is the oldest clock still working continuously.

Below: The Salisbury Cathedral clock as it was found, before being restored. It had been converted to pendulum.

THE WELLS CLOCK

The clock at Wells Cathedral, Somerset, is the second oldest in Britain, and in many ways the most interesting. It was installed about 1392 under the direction of Bishop Erghum, who had been transferred to Wells in 1388 from Salisbury Cathedral, Wiltshire, where he had been responsible for the clock installed in 1386. The two clocks are so similar in detail that it is virtually certain that they were made by the same hand. The Wells clock was dated incorrectly for many years, and fancifully attributed to a monk clockmaker called Peter Lightfoot. In fact, the clock may have been made by Johannes Lietuijt, or another of the group of clockmakers invited to England from Holland by King Edward III in 1368. The Salisbury clock just strikes the hours, and has no dial. The Wells clock strikes the quarters as well as the hours, has an astronomical dial and another dial added later, and three separate automata or automata groups. Striking of the Salisbury clock was controlled in a primitive manner so that the timing was not very exact, and attempts were made to remedy this. Precision of striking is obtained in the Wells clock by the ingenious technical development of a double lever. Also, the locking plate, by which striking is controlled, external on the Salisbury clock, was moved inside the frame of the Wells clock. Whether the quarter striking was introduced from the Continent or invented by the maker of the Wells clock is not known.

The original movement of the clock was removed in 1835, on the pretext that it was worn out – which it certainly was not, because it was later loaned to the Science

Right : The medieval knights on the Wells clock. They move in opposite directions, and here one is about to be felled.

Museum in London, where it is still working today. The second movement worked for 45 years until replaced by a flat-bed movement which continues to motivate the clock today. The second movement is still working, too, but in the parish church at Burnham-on-Sea.

The Wells astronomical dial is the finest still preserved in England. The main dial represents a day of 24 hours by the Roman numerals I to XII repeated, known as a double-XII dial. An image of the sun on a hand indicates the hour of the day. Inside this is a ring showing 1 to 60 on which a small star indicates the minutes. In the centre of the dial are two rotating discs, one over the other. The inner one has a moon painted on it, and the outer one a hole of the same size. They are geared ingeniously together, so that, as they rotate, they show not only the phase of the moon, but also its attitude in the sky. Rotation of the moon shows its age against a ring of numbers from 1 to 30, so it had to be adjusted monthly as a lunation is about $29\frac{1}{2}$ days. Opposite the moon is a medallion with an image of Phoebe (symbol of the moon), which is unique as it is counterbalanced so that Phoebe remains upright as she is rotated. In the centre of the whole dial is a ball surrounded by wavy carvings symbolizing clouds and sky.

At the corners of the dial are paintings of angels holding on to the four cardinal winds. The winds, represented by faces, are all blowing in the same direction, to 'turn' the Earth in its direction of rotation, opposite to the passage of the hand moving the sun. Thus the dial is a working model of the medieval concept of the universe, with the Earth and its four elements – earth, air, fire, and water – in the centre, surrounded by the first heavenly sphere of the moon, which rotates around it.

Above the astronomical dial are automata representing four knights. This automata group is contemporary with the clock. In the summer, the knights are now made to joust every quarter of an hour for the benefit of visitors, but they are given the assistance of an electric motor, installed in 1968. The hours and quarters are sounded by a much larger automaton or jack, known as Jack Blandifer, seated in a niche to the right, higher than the astronomical dial. He was made of oak, probably in the fourteenth century, and repainted in the seventeenth. The second dial is about a century later than the main one and is on the outside of the west tower of the cathedral. It shows the time on a 12-hour dial, and has above it two armoured

jacks who strike the quarter-hours on a small bell.

The jacks are about 1.25 metres (4 feet) tall; they are made of wood and are in armour of the latter part of the fifteenth century. Looking at this dial, only the quarters can be heard, as the hour bell is in the high central tower. The original clock was installed in the transept, and according to the late R. P. Howgrave-Graham, who did so much research on medieval clocks, it was unlikely that the public ever saw it. The clock was for the benefit of the cathedral officials and the hour bell was for the surrounding populace.

Above: The interior dial in Wells Cathedral. It shows the moon's phase in the centre at the correct angle and the model of Phoebe opposite the moon stays upright. Above the dial are the knights automata.

*Right : The exterior
dial of Wells Cathedral.
The wooden jacks above
it strike the quarters.*

THE HAMPTON COURT CLOCK

Two clocks are associated with the ill-fated Anne Boleyn, second of the English King Henry VIII's six wives. One is the Queen Anne Boleyn clock, reputed to have been a present to her from the King on their marriage in 1532. It is a wall clock on a bracket, both being of gilded metal. The case is surmounted by a lion bearing the arms of England, which are repeated on the sides. The two weights that hang below the bracket are engraved with the initials of Henry and Anne with true lovers' knots above and below them and the inscriptions 'The Most Happye' and 'Dieu et mon Droit'. The movement was made in the mid-sixteenth century but is not the original one. Queen Victoria bought the clock at the sale of Horace Walpole's possessions in 1842.

The second clock associated with Anne is the large astronomical clock made in 1540, after her death, and installed in the Anne Boleyn gate, facing east into the Clock Court at Hampton Court Palace. The colourful and informative astronomical dial and its mechanism have been restored, but an original bar inside bears the date 1540 and the initials N.O., those of Nicholas Oursian, also called Urseau and Wourston, a French clockmaker, one of a number of French and German craftsmen employed by Henry, who had much interest in the arts and particularly in astronomy. Oursian was appointed keeper of the clock in 1541 for a wage of fourpence a day, so it is likely that he made the dialwork that he signed. (There is no evidence, but Nicholas Urseau, Clockmaker to Queen Elizabeth I, may have been the son of the keeper of the Anne Boleyn clock.) It was more likely to have been designed, however, by Nicholas Cratzer, who came from Munich with a reputation for learning in astronomy and mathematics to lecture at Oxford (although he never learned to speak English) and to be tutor to Sir Thomas More's children. Henry appointed him to the post of royal astronomer.

In 1649, when Hampton Palace was retained for the use of the Lord Protector, Oliver Cromwell, the dialwork was used also to drive an orthodox dial installed on the other side of the gateway. The original movement would have had verge and foliot control and this was no doubt converted to pendulum during Cromwell's time. He had a known interest in the invention of the pendulum. In 1835 the original movement was removed and disappeared. It was replaced by a clock from St James's, which worked until 1880, but

was never linked to the astronomical dial. Then, in 1879, a Gillet and Bland flat-bed striking and chiming movement was installed with an arrangement by which it provided impulses every 15 seconds to drive the astronomical work. Both movement and dial were last restored in 1959–60 by Thwaites and Reed.

The dial is 3 metres (10 feet) in diameter. There is a long pointer bearing an image of the sun which turns once in 24 hours and indicates the time on the outer double-XII dial. The month, the date and the position of the sun in the zodiac can also be read by the

Left : A lantern clock on a bracket which tradition holds was given to Anne Boleyn by Henry VIII. It is now in the library at Windsor Castle.

Right : The 'modern' flat-bed movement of 1879 which operates the dials at Hampton Court Palace. Behind the pendulum rod on the left is the three-legged Grimthorpe gravity escapement, in the centre, the remontoire, and right, part of the fly to slow the striking. Weights on pegs, bottom left, regulate the pendulum.

Right : The Hampton Court movement. The going train in the middle has an unusual drive straight down to the dial, where the motion work is. The count wheels with notches are for hours on the left and quarters on the right.

same hand from the next rings of information. These rings are on a separate dial that revolves once in a sidereal day (a rotation of the Earth measured by a star instead of the sun), which is 3 minutes 56 seconds shorter than a mean solar day (a day as shown by a clock, not a sundial). This dial therefore goes a little faster than the hour hand with the sun image, that is the hand moves anti-clockwise in relation to the dial, and for this reason the dates and months are marked anti-clockwise. The sun pointer is attached to a dial that is marked from 1 to 29½ and turns once in a mean solar day. The innermost dial is the lunar dial, which has an image of the moon and a wedge-shaped pointer indicating the moon's age in days on a 1 to 29½ scale. The image of the moon moves behind a hole to represent the correct phase. There is one more practical indication on this ingenious dial. A small red pointer by the inner end of the sun pointer gives a reading of the moon's southing on the inner double-XII dial. The southing is when the moon crosses the meridian, from which the times of the tides (which are caused by the gravitational pull of the moon) can be calculated. For example, high water at London Bridge is about 75 minutes after the moon's southing.

Right : The double-XII dial at Hampton Court Palace. The image of the sun, approaching XII, turns once in 24 hours.

THE FIRST PENDULUM CLOCKS

The man who had most influence on the development of accurate timekeeping was undoubtedly the Dutch astronomer and physicist Christiaan Huygens. He invented the pendulum clock and, 18 years later in 1675, the spiral balance spring (hairspring) for watches.

From a letter he wrote, it is known that Huygens spent Christmas Day of 1656 making his first model of the pendulum clock. Nearly six months later, he granted the right of making the clocks to a clockmaker of The Hague. The first of these, a small wall clock, carries a brass plate on the dial, marked by the man who made the clock, 'Salomon Coster Haghe met privilege 1657'. It is now in the National Museum of the History of Science, at Leiden in Holland.

Huygens hung his pendulum from two threads, and linked the pendulum to the clock escapement, a traditional verge and crown wheel, by a device called a crutch, which became universal. The 13.7-centimetre (5½-inch) pendulum had curved pieces called cheeks each side of the suspension threads, because Huygens had already discovered that a pendulum that swung in a circular arc was not isochronous (that is, it did not swing at the same intervals in dif-

ferent degrees of arc) and that to be so it had to swing in a different sort of curve that was steeper at the ends, called a cycloid. The curved cheeks were supposed to achieve this. It was found later, after the invention in about 1671 of the anchor escapement, which released the real potential of the pendulum, that it is much easier to keep the arc of swing as constant as possible. The clock was a typical *Haage clokje* of the time, in a rectangular-framed case with a brass dial on a black velvet background. It was driven by a spring, although Huygens's book of 1658 showed a clock that was weight-driven, using an endless rope holding the weight to maintain driving power while the clock was being wound, an idea that was applied to large numbers of clocks subsequently and is commonly used today for the automatic winding gear of public clocks. Realizing that the wide pendulum swing of his first design would cause errors, he tried to limit the swing of the pendulum by gearing, which was not practical and was probably, as far as is known, never applied to a clock.

Although the pendulum clock was invented by a Dutchman, it was first and most thoroughly developed in England, by Ahasuerus Fromanteel, who was third-generation

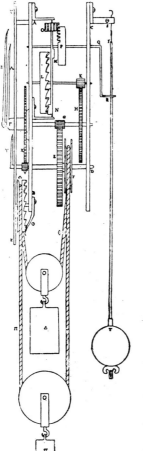

Left : Huygens's illustration in 1658 of his first pendulum clock. The gearing at the top was introduced to reduce the arc of the pendulum. It was impractical.

Right : The design adopted by Huygens for his first pendulum clock, made by Coster. The gearing between crown wheel and crutch is eliminated by turning the crown wheel. The sketch was in Huygens's book of 1673.

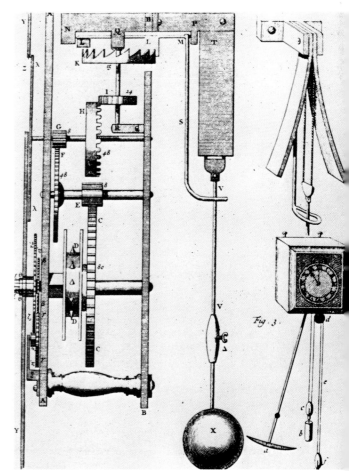

Above : The dial of the original pendulum clock of 1657, which has a velvet ground. It is a small spring-driven wall clock.

THE BATH TOMPION

Thomas Tompion is probably the most famous of all English makers of clocks and watches. Born in 1639 in Ickwell Green, Northill, north Bedfordshire, he was trained as a blacksmith but went to London to set up business as a clockmaker at The Dial and Three Crowns, Water Lane, Fleet Street, London. Dr Robert Hooke and the first Astronomer Royal, John Flamsteed, soon began to take advantage of Tompion's skills and aptitude. He made the first clocks for the Greenwich Observatory, the first equation of time clock, now in the Royal Collection, and the so-called 'Record Tompion' (because it fetched a record price in an auction), made for King William III which was in the upper middle room of the Palace at Williamsburg, the restored eighteenth-century town in Virginia, but was removed in 1978.

In 1709, four years before he died, Tompion presented a magnificent longcase equation clock to the Pump Room of the Spa in the City of Bath, presumably as a recognition of treatment he received for some rheumatic ailment. The clock stood in the original Pump Room, but more or less where it is now. It is tall, over 3 metres (10 feet)

English despite his name, and made clocks in Mosses Alley, Southwark, London. One of the family, John, was an apprentice to Coster at the time that Coster was granted the patent, and transmitted details to Fromanteel, undoubtedly with the approval of Huygens. Fromanteel not only made some of the first pendulum clocks, but also appears to have introduced the longcase clock, later called the grandfather clock, but his first longcase clocks had pendulums only about 25 centimetres (10 inches) long. After the anchor escapement was introduced, the pendulum was almost always about 1 metre ($3\frac{1}{4}$ feet) long, beating seconds.

As fundamental to the history of timekeeping was Huygens's balance spring, the invention of which was also claimed by the English scientist Dr Robert Hooke, although his appears to have been a flat spring and was applied to a pendulum, which missed the point of replacing the force of gravity. The balance spring was and still is used in very large numbers of clocks as well as watches. Huygens also spent many years in attempting to make a timekeeper that would 'solve the longitude' to aid navigation at sea, but without success. This was to come later after many experiments at sea.

Right : The sundial originally supplied by Tompion to check the clock he presented to the Spa authorities at Bath, Somerset.

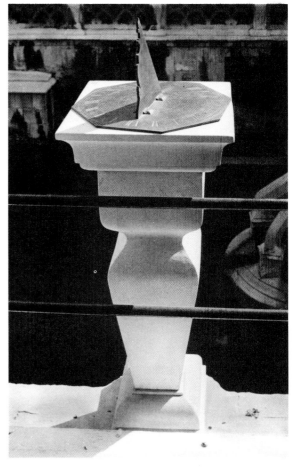

Above : Thomas Tompion, after an engraving by G. Kneller.

Right : The longcase clock by Thomas Tompion in the Pump Room at Bath. In the dial arch is a dial showing the equation of time, so that the mean time of the clock could be checked against the sundial's solar time.

high. The case is of solid oak, with an unusual case trunk, shaped in the front like a column, but otherwise quite plain compared with many special clocks of the time. The movement, unlike those of almost all other long-case clocks, does not strike. It has, however, the usual seconds pendulum about 1 metre ($3\frac{1}{4}$ feet) long, which looks shorter than usual because of the height of the case. The movement is simple despite the equation work and the fact that it runs for a month at a winding. There is only one weight, of course, as there is no striking, and there is maintaining power to keep the clock going while it is being wound. There is a shutter over the winding hole, and when a lever is moved to displace this, it also causes a spring-loaded pawl to push on a tooth of the centre wheel to keep the clock working.

The equation dial in the arch above the time of day dial shows how fast or slow the sundial should be compared with the clock. Solar time varies throughout the year because the lengths of the days vary. This is caused by the facts that the Earth moves faster in its orbit when it is nearer the sun in January and that the tilt of the Earth's axis alters the sun's apparent movement. Clocks average out the days and the old name for

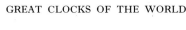

THE ROSKILDE CLOCK

A clock with interesting and noisy automata can be found in Roskilde Cathedral, in Denmark, where St George slays the dragon. At the hour, the saint rises in his stirrups and strikes a blow of his sword for every hour, while his horse gallops on the poor dragon, which makes agonized noises. To the right of this scene is a bell, and then two jacks in medieval costume. The larger one is called Peder Dove or Per Dover (Peter the Deaf) and strikes the hours with a hammer. His companion is Kirsten Kimer (Christine the Ringer), who strikes the half-hours and the quarters. In the eighteenth century, the clergy had St George put out of action be-cause he was more interesting than their services, but he was put back to dragon-slaying in 1884, when the clock was restored, although he had a period of inaction again and for the same reason some years later. The dial below the automata is a double-XII and is unusual because there is only one hand and that has a sun at one end and a moon at the other. The present clock replaced an earlier one, made by Peter Mathieson, of Copen-hagen, which was installed in 1741. The restoration of 1884 was carried out by Bertram-Larsen, of Copenhagen.

Right : The equation dial shows a plus or minus figure of minutes and seconds to be subtracted from or added to sundial time.

Right : Part of the birdcage movement of the Roskilde Cathedral clock. At top right is the pin-wheel of the escapement, often used on the Continent with a pendulum instead of an anchor escapement.

Far right : The medieval wooden figures of St George and the Dragon, which perform above the Roskilde clock dial.

time shown by clocks was equal hours. To set a clock, then, an accurate sundial was needed and either a table giving the equation figures or an automatic indication such as is provided by an equation clock. Tompion supplied a sundial, and recently this was rediscovered and presented to the Pump Room.

In the Bath Tompion, the equation hand moves back and forth over a dial divided into minutes. It is operated by a cam which rotates once a year, being turned by a two-stage worm gear driven from the motion work that provides the drive for the hour hand. A lever with a roller on the end is kept in contact with the edge of the kidney-shaped cam by a counterweight, which also tensions the chain (the same type as used in fusee watches) that turns the hand, which has a little image of the sun at its tip. As in most orthodox longcase clocks, there is a date indication aperture in the dial, that has to be set monthly, but there is also an annual calendar that presents the month and date in an aperture within the equation dial. As a yearly motion was needed for the equation work, it is not surprising that an annual calendar was included, but it is odd that the ordinary date indication was retained. Another curiosity is the oval-section brass pendulum rod.

BREGUET'S *PENDULES SYMPATHIQUES*

Abraham-Louis Breguet, a Swiss who worked in Paris except for a period during the French Revolution, is perhaps the most renowned maker of all time for the excellence of his work and his outstanding solutions to technical problems. The most dramatic, although not the most useful, of his productions were his *pendules sympathiques*. Such a clock was designed to keep on time a special pocket watch that was worn during the day. In the top of the clock is a holder. 'Then,' as Breguet wrote to his son in 1795, 'every night on going to bed, you put the watch into the clock. In the morning, or one hour later, it will be exactly to time with the clock. It is not even necessary to open the watch. There will be nothing visible externally to show where it has been touched. I expect from this the greatest promotion of our fame and fortune.' Breguet claimed that it took only one day each to modify a suitable clock and pocket watch, and this may have been so for the first models which just set the minute hand of the watch on the hour, an idea also used later in the first public time services, which used an electric current to centralize the minute hands of clocks in public places and offices subscribing to the service. But Breguet later made clocks that wound the watch as well as setting the hand within limits; then clocks that set both hands regardless of where they were; and also clocks that *regulated* the watch as well, according to how much it was gaining or losing. This is one of the earliest instances of the use of the feedback principle of control. In practice, it does not work well with watches, because their timekeeping does not improve with continuous regulation; timekeeping is better if regulation is spaced over much longer intervals. The regulation mechanism of the *pendule sympathique* will work within 20 minutes fast or slow, but if a well-regulated watch that is this much out is placed in the clock, the rate will be upset. So Breguet was strictly correct in thinking of the *pendules* mainly as promotions for his business. In 1814 the Prince Regent of England bought one, which is now in the Royal Collection.

The most successful of the *pendules* was completed 13 years after Breguet's death, but may have been started in his time because an earlier equally complicated *pendule* took 20 years to make. This one, the last, is about 40 centimetres (15½ inches) high, has a dead-beat Graham escapement, strikes the hours and quarters, and runs for eight days at a winding. (The earliest *pendule* had to be wound daily.) The watch is a half-quarter repeater, so that, when operated, it will strike the hour just gone on a lower note, the quarter as one, two or three ting-tangs, and then either nothing or the higher note to indicate whether or not the last 7½ minutes has passed. It has a hole at the edge of the bottom of the case to take the winding square from the clock and, by the hole, two small pins. The watch is placed in a cradle on top of the clock, and, at 3 a.m., the hands of the watch will be set to time regardless of what they showed. The watch will also be wound until an indicator on the watch indicates that winding is complete, when the watch switches off the winding mechanism. There is no regulation. Correction of the hands is startling to the uninitiated because

Below: One of Breguet's pendules sympathiques. *It was bought by Albert, the Prince Regent, in 1841, and is in the Royal Collection.*

they are returned to show 3 o'clock by the shortest route and may turn in opposite directions. Power for winding the watch comes from the clock's striking train. The clock switches on the power by a complication of levers, pins and wheels, and the watch switches it off by means of one of the pins in the side of the case. The other pin sets the hands to time by uprighting a knob on the cannon pinion which carries the minute hand and another on the hour wheel, which sets the hour hand to 3. The arrangement is broadly similar to that used on timers for returning hands to zero. George Daniels, the present-day maker on whose shoulders the mantle of Breguet has fallen because he makes comparable watches by hand, regards the *pendules* as fine works of art for their own sake, jewels of misplaced ingenuity.

BIG BEN AND GREAT TOM

The original clock at St Paul's Cathedral in London had a bedpost frame (like a four-poster bed) and was made about 1706 by Langley Bradley, a famous maker who was Master of the Worshipful Company of Clockmakers from 1726 to 1738 and who also made the clock for St Giles Cathedral in Edinburgh. It was built in the design that had evolved over centuries, with the driving barrels, carrying the lines and weights, at the bottom of the frame and the trains of gears vertically above them, the going train for turning the hands in the middle, the striking train on the left and the quarter-striking train on the right.

After Edmund Beckett Denison (later Lord Grimthorpe) bulldozed his way through all opposition to impose his own designs on the Westminster clock, and the clock had been built and was going much better than anyone (except perhaps Denison) had expected, there was still opposition from the reactionary section of the clockmaking trade to the 'much vaunted new machine at Westminster'. Some insisted that Bradley's old clock went better 'despite it [Big Ben] possessing an escapement invented by amateurs who consider themselves the depositories of all horological knowledge'. In fact, Denison's design and the performance of the clock were much better than those of the Bradley clock.

The design, an idea imported from the Continent, is known as flat-bed because the frame of the clock is a rectangular cast-iron frame, two long girders and two short ones making a rectangle. (Today rolled steel girders are sometimes used.) The arbors carrying the barrels and wheels are positioned along the frame and across the narrow width, the

Right : Edmund Beckett Denison, later Lord Grimthorpe, designer of the great Westminster Palace clock 'Big Ben'.

Below : One of the sketches of the Westminster clock. Note the cam barrels for operating the hammer ropes of Big Ben on the left, and the quarter bells on the right.

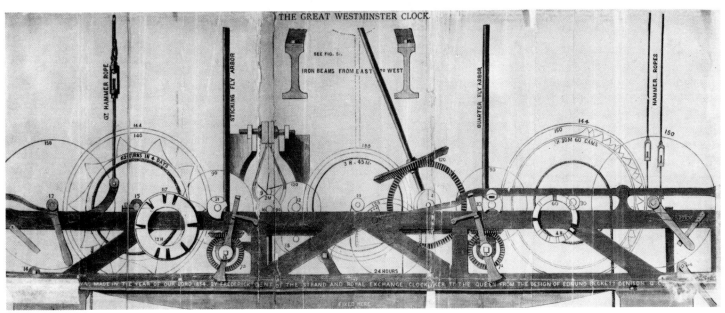

bearings being separate pieces bolted to the top of the frame to enable any part to be removed easily without dismantling the frame or any other part of the clock. The double three-legged gravity escapement gave the clock its superior timekeeping. In most escapements, the impulse that keeps the pendulum swinging is applied by the driving weight or spring through gearing. It can vary with inaccuracies of the wheels and other parts, with varying friction, and, if spring-driven, with variations in driving force. The gravity escapement applies a constant measured impulse by using the power of the driving weight to raise a light arm. The weight of the arm provides the impulse as it rests on the side of the pendulum rod. It is lifted out of the way as the pendulum swings back, then released again. Many attempts have been made in the past to design gravity escapements, with varying success, and Denison's has been the most successful for large mechanical clocks from its first application until today.

Denison, with the assistance of the clock-maker Frederick Dent, fitted an experimental gravity escapement to a clock that was submitted to the Astronomer Royal, George Biddell Airy, for testing at the Greenwich

Far left : Probably the best known clock in the world, in the tower of the Palace of Westminster, home of the Houses of Parliament, is known by the name of its hour bell, 'Big Ben'. The 6.86m (22 ft 6 ins) diameter dials are 55m (180 ft) above ground, and the minute hand weighs 102kg (2 cwt) and is 3.36m (11 ft) long.

Left : Old pennies on a platform above the two-seconds pendulum raise the centre of oscillation and speed up the pendulum.

Left : Part of the flat-bed movement of 'Big Ben'. The beam across the top carries leading-off work to operate the hands, and brackets for the huge flies.

Observatory. It performed so well that Airy, formerly strongly against all forms of gravity escapement, changed his mind and approved of its incorporation in the Westminster clock, although further development was still needed. At this time Denison was involved with a clock Dent was making for Fredericton, in New Brunswick, Canada. A gravity escapement was fitted to this. The invention was not protected, so, as Denison wrote later, 'as it is not patented, it may be made by anybody'. In fact, according to T. R. Robinson, Denison would become enraged with any clockmaker who would not adopt his escapement. Big Ben's pendulum beats (swings from one side to the other) two seconds, so the hands jump every two seconds. There is a great weight of mechanism to jump, day in, day out, without ceasing for centuries. Each of the four minute hands is 4.5 metres (15 feet) long and about 100 kilograms (2 hundredweight) in weight, and the four hour hands are 2.7 metres (9 feet) long and proportionately heavy. Although the hands showed accurate time, the striking was not at first accurate. The jumps of the going train were too small to provide accurate release of the striking. The problem was overcome by an earlier release of the striking train from its locking, to be held up again by a second locking catch, which is released by a cam turning once in 15 minutes, instead of the once-an-hour rotation of the normal let-off cam. This happens at the 58th second, because it takes two seconds for the striking train to gather momentum to drop the previously raised hammer on Big Ben exactly on the hour. The timekeeping of this 5 metres (16 feet) by 1.7 metres (5½ feet) clock movement was well within the limits orthodox clockmakers declared impossible. On most days it was less than 0.2 of a second in error. It was, and still is, an independent timekeeper, although it had a galvanic link with the Royal Observatory at Greenwich to provide a time check. The chimes and striking of the clock were for many years broadcast in full before the news bulletins on the radio, and indeed played a very important part in sustaining morale in occupied Europe during the Second World War. Surprisingly, the British Broadcasting Corporation did it much less than justice on its centenary in 1959 with a largely irrelevant programme, 'The King of Clocks', containing a number of factual errors.

The clock and its tower and bells survived two world wars, yet nearly came to a disastrous end at 3.45 a.m. on 5 August 1976. A shaft

Left : An eighteenth-century engraving of St Paul's Cathedral showing the old clock in the right hand spire.

Below : The old bell at St Paul's is known as Great Tom and was recast from the original at Westminster, on which the Cathedral clock strikes the hours.

failed and the 1.5 tonne chiming weight fell down the 40 metres (130 feet) tower. The reaction caused the 760 kilograms (15 hundredweight) barrel to jump out of its bearings on the frame and across the clockroom, breaking the frame. All associated wheels were ripped out and the main wheel was broken into several pieces. The clockroom looked as if a bomb had hit it. Investigation showed that the fly shaft (driving a fan brake to space the chiming) that broke was a length of gas piping (presumably a replacement) and that there were cracks in other parts. Thwaites and Reed, who looked after the clock, rebuilt the chiming side so that the new work is in keeping with the original, although they had to provide an extra pillar to support the frame. The clock was set going again on 4 May 1977.

The casting of the hour bell, the true Big Ben, caused almost as much trouble as the clock itself. This was because Denison claimed to be an expert on bells and had a contempt for bell-founders. Warner of Cripplegate were awarded the first contract after a specification had been drawn up by Denison and sent to three bell-founders. It was for a 14 tonne bell, one of the biggest ever to be cast in the country. Sir Charles Barry, the Westminster architect, had specified this weight of bell, but had neglected to estimate its size and to design the tower so that it could be installed or even hauled up. Denison had therefore to modify the traditional bell form to fit the tower, and the bell-founders approved of neither the new shape nor the composition. The first casting in 1856, in a borrowed foundry in Stockton-on-Tees, produced a bell that was nearly 2 tonnes overweight and half a note out; it was intended to

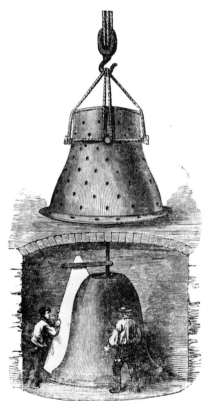

Above : Wenceslaus Hollar's engraving of the New Palace Yard. The stone tower on the right, opposite Westminster Hall, built in the early fourteenth century, had a clock that struck the hour on Great Tom.

Left : Making the core of a bell mould of sand, using a template.

sound E flat. It was accepted, however, shipped by sea, and hung in a frame at Westminster for testing. The original 355 kilograms (7 hundredweight) hammer produced little sound, so one of 660 kilograms (13 hundredweight) was substituted. It had to be raised by ten men, and the bell, after it had been struck a few times, cracked. Two years later, while the arguments were still raging, George Mears, of the famous Whitechapel Foundry in London, agreed to recast it to a modified shape. The result was a 13.8 tonne bell that could be hauled up the tower, but the striking train had to be modified to

Left : Unshipping the first Big Ben, which had been transported from Stockton-on-Tees to a wharf in Lambeth on the Thames.

Below : The second Big Ben, cast in Whitechapel, 'arriving in triumph' at Westminster, drawn by 16 horses, as shown in The Illustrated London News *at the time (1858).*

strike it. The clock itself was started in May 1859: the striking followed in July, and the quarter striking in the following September. During a Parliamentary debate after the hour bell was installed the overweight Chief Commissioner, Sir Benjamin Hall, made a long speech calling for a dignified name for the bell and a backbencher called out 'Call it Big Ben and have done with it!' After a short time in service, Big Ben was found to be cracked and was silenced for three years, during which time the hour was sounded on the largest quarter bell. Then, with Denison protesting that it should be recast, the bell was given an eighth of a turn and a lighter hammer fitted. It worked and is still working today.

The St Paul's authorities were very impressed with the new Westminster clock, and the Dean approached Denison to design one for St Paul's. The clock, after it was installed in 1893, was claimed to be the biggest in the United Kingdom. It was slightly longer than the Big Ben clock. The movements are very similar, the main difference being that the pendulum at St Paul's hangs from the clock frame, whereas that at Westminster hangs from the wall behind it. The escape wheels turn in opposite directions. Chiming is also different. St Paul's is a traditional ting-tang on two bells, whereas the Westminster clock plays the Cambridge quarters on four bells; this was adapted from a phrase in Handel's *Messiah* for Great St Mary's Church, Cambridge, but is now generally known as the 'Westminster chime'. The weights of both clock trains are the same – 1.5 tonnes each for the striking and chiming trains, and only 75 kilograms (1½ hundredweight) for the going trains. The St Paul's clock strikes on a 5 tonne hour bell known as Great Tom, of which Denison scathingly remarked, it 'might be recast for nothing into a very much more powerful and infinitely better bell (for it could hardly be worse) by reducing the weight. . .'.

Westminster had a clock before Big Ben. The old Westminster clock was in an early-fourteenth-century clock tower in the Palace Yard, only a short distance from the present clock tower. First reference to the clock was in 1286, and it is believed to have worked for about 300 years, as various royal clockmakers were charged with looking after it. The hour bell, named Edward after King Edward I, was to be the hour bell for Langley Bradley's St Paul's clock, but fell off its carriage in what subsequently became known as Bell Yard, in The Strand, London, was cracked, and had to be recast.

THE MOST COMPLICATED WATCH

A very rich Portuguese named A. A. de Carvalho Monteiro in 1897 commissioned the firm of Le Roy, Paris, to make a watch that, when it was completed four years later, was probably the most complicated ever made, and remains so. Le Roy's had a problem of delivery, but it happened that the King of Portugal, who was a client, called on their workshops in Paris and offered to take the watch to his palace in Lisbon to await collection. It remained in Lisbon until 1953, when it had become the property of a jeweller. That year, a public subscription was organized to buy it and present it to the Museum of Fine Arts in Besançon, the centre of France's watchmaking industry. The heavily embossed 18-carat gold case is 7 centimetres (2¾ inches) in diameter, and weighs 228 grams (8 ounces). The movement has a lever escapement with a temperature-compensated balance. There are two dials, one on the front and one on the back, and the watch provides:–

time of day
day of the week
date
month, regardless of length

Right : One side of a watch believed to be the most complicated ever made, by Le Roy, Paris, for a rich Portuguese customer.

year, including leap years for 100 years
phases of the moon
seasons, solstices, and equinoxes
equation of time
elapsed time by chronograph with zero-setting mechanism
minutes and hours of elapsed time
how much the watch is wound (up-and-down indication)
grande sonnerie striking, with silencing mechanism
repetition of the hour, quarter, and minutes on three bells, when required
state of the sky in the northern hemisphere on the day indicated on the calendar, with a sky and horizon for Paris with 236 stars and the same for Lisbon with 560 stars
state of the sky in the southern hemisphere (by fitting a replacement mechanism to change the direction to from west to east), with a sky and horizon for Rio de Janeiro with 611 stars
time of day in 125 cities around the world
sunrise and sunset times in Lisbon
temperature by a centigrade thermometer
humidity by a hair hygrometer
height by an altimeter registering up to 5000 metres
direction by a compass

Despite all the other things the watch has to do, the chronograph (a timer) has a positive action better than many others of the period, and the repeater mechanism has a silent-running train of gears. There is even a means of regulating the watch without having to touch the movement.

OBSERVATORY CLOCKS

The invention of the pendulum clock was a boon to astronomers, who could determine time by an observation, but not preserve it accurately. Timekeeping precision was first achieved by accuracy of workmanship, then by understanding the nature of errors and dealing with their causes. The first to come under intense scrutiny was the effect of varying temperature on the length of the pendulum rod, something not even thought of until the early eighteenth century.

The first clockmaker to measure the contraction and expansion of metals was John

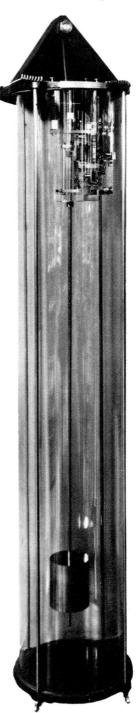

Right : The other side of the watch reputed to be the most complicated in the world. It was so interesting to the king of Portugal that he offered to deliver it.

Left and right : The Shortt free pendulum in its low pressure chamber on the left and the slave pendulum with which it is associated on the right. This is a converted Synchronome clock.

Harrison, maker of the first timekeeper to go accurately at sea, who devised a temperature-compensated gridiron pendulum. George Graham followed with a mercurial pendulum for precision regulators. The problem was largely solved by Charles-Edouard Guillaume's invention of invar, a metal that did not vary in length to any appreciable extent in different temperatures. Attention was also being paid to the effect of the escapement itself on timekeeping. Graham invented the dead-beat escapement, the German Sigmund Riefler devised an accurate escapement that acted through the suspension spring, and

another Englishman, W. H. Shortt, produced the ultimate in electro-mechanical clocks, the free pendulum clock.

The free pendulum clock has two pendulums, each impulsed by basically simple electrically reset gravity escapements and linked together by what would today be called a positive feedback system. One, the free or master pendulum, is relieved of all work; it has nothing to do except swing as accurately as possible. Like other precision clocks, it is enclosed in a partial vacuum chamber, to eliminate errors caused by barometric changes, kept at constant temperature and constant arc of swing, as well as being situated where there is the minimum of disturbance by vibration, even by distant earthquake. It is kept swinging by an impulse every minute from a light arm with a D-section pallet on

Above : Top of the free pendulum, which has no connection to any parts of the rest of the mechanism. It is impulsed by a jewel on a light arm which is released by the slave pendulum. The jewel drops on a tiny wheel mounted on the free pendulum rod, designed to give impulse only at the correct instant.

Right : The hit-and-miss synchroniser on the slave pendulum, which runs slightly slow. When the impulse lever of the free pendulum is restored electromagnetically, the current also pulls down the lever above the coil on the left. If the slave pendulum is slow, the lever will engage with the vertical leaf spring and speed up the pendulum. Normally the hitting and missing occurs on alternate half minutes.

the end that is dropped on to a tiny brass wheel on a short arm projecting from the pendulum rod. The arrangement provides a precise but gradual impulse at the same position in any swing. The other pendulum, the slave, does any work that is necessary. It is a seconds pendulum, like its master, and is impulsed by a simple gravity arm in the same way, except that the escapement is more robust. In fact, the slave unit is a commercial Synchronome master clock. A small spring pawl attached to the slave pendulum rod turns a 30-toothed count wheel one tooth at every other swing. Attached to the wheel is a vane that trips a latch every minute to release the gravity arm and impulse the slave pendulum. When the arm has done its work, it drops on to a switch, which energizes an electromagnet and relatches the gravity arm. The current that does so also releases the gravity arm of the free pendulum. It has a third function, to advance the hands of the slave dials. When the free pendulum has been impulsed, its gravity arm also makes an electrical contact and it is reset. The current that does this goes through the coil of a device on the slave pendulum called a hit and miss synchronizer. The coil pulls a lever momentarily into the path of a spring strip attached to the pen-

dulum rod. If the pendulums are swinging in time, the lever misses the spring, but if the slave is slow, the lever makes contact with the spring to bend it and give the slave an extra small impulse. The slave is regulated to lose six seconds a day compared with the master, which means that the spring is usually engaged at every other swing. The current that operates the synchronizer also advances the hands of a master dial. What happens, then, is that the free pendulum determines the time of the whole system, because of the geometry of its gravity impulsing system, and it keeps the slave in step with itself.

W. H. Shortt set his first clock going in Edinburgh Observatory at Christmas 1921, and its first year's performance created a horological sensation. No. 3 Shortt-Synchronome at Greenwich did not vary by more than a hundredth of a second in its first year, and No. 41 at the Washington, D.C. Naval Observatory kept within a fiftieth of a second while running continuously for three years. Performances could be even better, as these variations seem to be caused by sudden very small and unpredictable dimensional changes in the invar pendulum rods.

For short-time accuracy, the mechanical clock was surpassed by the quartz crystal clock, the first precision version of which was designed by W. A. Marrison, a Canadian working for the Bell Laboratories in the United States, in 1929. An accurately machined and correctly shaped piece of quartz (rock crystal) is kept in physical vibration (by reason of its piezo-electrical properties) in an oscillating current of one kilocycle a second. If the quartz crystal is kept hermetically sealed in a vacuum chamber at constant temperature, like a master pendulum, its vibrations will remain exceptionally constant and will control the oscillating circuit, which can be used to run a motor to turn clock hands. The first observatory quartz clocks were large units several cubic metres in volume and kept time to the equivalent of one second in 30 years. Quartz suffers from an ageing effect, however, which causes slight drifts in timekeeping, so it was usual to employ quartz clocks in groups and average the times they provided. Quartz timekeepers have been revolutionized by the transistor and microcircuit so that some can be contained in small wrist-watches.

The next big step in timekeeping came with the caesium atom clock, developed by Louis Essen and J. V. L. Parry, of the National Physical Laboratory. They used the vibrations of caesium atoms at 9,192,631,770

Left : The original atomic clock at the National Physical Laboratory. It kept time to the equivalent of one second in three hundred years. The latest version is more than ten times more accurate.

cycles a second, to monitor a quartz clock to keep it to even more precise time, equivalent to one second in 3000 years, or one millionth of a second a day, more accurate even than the astronomical observations by which time is determined. The atomic clock detected that the Earth was slowing down. The 'clock' itself comprises an 80°C (180°F) oven containing a small quantity of caesium, from which atoms stream at a velocity of about 20,000 centimetres (656 feet) a second. The beam passes through the field of a magnet, then through the elements of a radio oscillator, which resonates at the frequency of caesium, and another magnetic field, until it reaches a detector, a hot tungsten filament. In the centre of the oscillating circuit elements is a focusing slit. If the oscillator is not working, the beam is dispersed by the magnetic field,

but if it is on and at the correct frequency and strength, some of the atoms are deflected in the opposite direction and are focused on the detector. The signals from the detector are stronger the more accurate the frequency and can be fed back to keep the frequency from drifting. The oscillator is used to calibrate or control the quartz clock or clocks. The atomic clock works because if an atom has an electron in its outer shell, the electron can spin in one direction and the nucleus in the other, or both can spin in the same direction. The two different states cause different magnetic fields, by which means beams can be focused. Spin can be changed from one direction to the other, which causes either emission or absorption of energy, and this permits the interaction between the beam of atoms and the radio oscillator of the atomic clock.

JENS OLSEN'S CLOCK

Large astronomical clocks built in the twentieth century are uncommon. Jens Olsen's clock in Copenhagen Town Hall is unique. Olsen (1872–1945) was apprenticed to a locksmith and, as a journeyman, travelled in Germany, then went to Basle, Paris and London. He had begun to teach himself clockmaking and was particularly interested in gear train calculations. There is a story that, as part of the Strasbourg clock mechanism was being repaired, he hid himself under a tarpaulin to study the gear train. He returned to Denmark to work as a clockmaker for a firm and also to set up on his own making clocks. His interest in astronomy developed so much that he was soon calling himself an 'astro-mechanic'. He had not lost sight of his intention to build a magnificent

astronomical clock, and as a preliminary he made a watch showing mean and sidereal time and, with Professor Stromgren, the astronomer, acquired a patent for the mechanism. He assigned the patent to a Swiss firm. With his proceeds, Olsen built a private observatory. When he was about 50 years old, he had completed his calculations for the clock and submitted them to Professor Stromgren, who commended them with a comment about Olsen's remarkable combination of mechanical and mathematical gifts. The problem was how to raise capital. First attempts were fruitless. Some time later, the Clockmakers Company set up a committee to help, but for many years that was unsuccessful. Under a new and dynamic chairman, and during the German occupation, the Employers' Union made a grant of

Below : The dials of Jens Olsen's clock in Copenhagen, which is in three main sections. The most recent of the great astronomical clocks, it was completed in 1955.

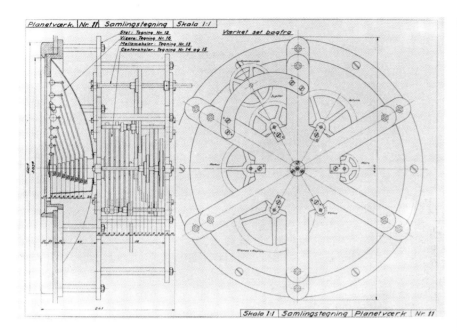

Above : One of the working drawings made by Olsen himself when designing his astronomical clock.

Right : Jens Olsen working on his clock. He died before it was finished, but a team of co-workers completed it and the whole population of Denmark subscribed to the cost as a symbol of liberation from the Nazi occupation.

100,000 kroner. This was not enough, but stimulated so much interest that the newspapers appealed to the whole population of Denmark to subscribe as a symbol of the nation's unity under occupation. The response was overwhelming, and, at the age of nearly 70, Olsen was able to set to work. First he was asked to make sure that the work could be completed if he died meanwhile, so he prepared six sets of drawings, each set comprising 46 main plans and 214 detailed ones. In 1944 the Technological Institute put a workshop at his disposal and he started making the clock with the help of his son. Unfortunately, his son was killed by a German bullet during a general strike, and Olsen had to carry on with the help of other clockmakers. Then, on 17 December 1945, he too died. But the clock was finished by his helpers. The team that started with Olsen and then finished the clock comprised an architect, two clockmakers, four instrument mechanics and two engravers.

The clock draws large numbers of visitors today. The indications are claimed to be more accurate than those of previous astronomical clocks. There are 11 self-contained mechanisms, linked to each other by trains of gears, arbors or steel ribbons translating crank motions to differential gears. Ten have dials and the eleventh has none and is the controlling unit. The dials are in three sections. The centre section has a mean time dial, a sidereal time dial, and the main calendar, which shows the Dominical Letter, the Epact, the solar cycle, the Indication and the lunar cycle on smaller dials, and the following in digital form: the month, date, day of the week, phase of the moon, and all movable

feasts, which are altered by the clock over a period of about six minutes from 24.00 hours on New Year's Eve. Movable feasts vary yearly and it takes that time for the clock to compute them mechanically. The feasts are Shrove Tuesday, Maundy Thursday, Good Friday, Easter Sunday, Easter Monday, All Saints' Day, Ascension Day, Whit Sunday and Whit Monday. The left-hand group of three dials shows local time, true solar time, the equation of time, world time, and sunrise and sunset by mean time and also true solar time, and is accompanied by a Gregorian calendar. The right-hand group has a similar layout and shows a star map, the motion of the celestial pole, the geocentric revolution and the heliocentric revolution, with a calendar of the Julian year and day numbers. The clock is contained in a glass case in a room air-conditioned by a special ventilating plant. It is controlled by a pendulum with a gravity escapement based on that invented by E. B. Denison for the Westminster clock, Big Ben. Jens Olsen's clock is not as large or to the layman as dramatic as the classical astronomical clocks of the past, but to the clockmaker, the movements that are visible behind the dials are a supreme example of the clockmaker's art and mystery.

Glossary

ACCUTRON The first commercial electronic watch, controlled by a tuning fork and introduced in 1960.

ALTITUDE DIAL A sundial depending on the height of the sun in the sky, i.e. the length of the shadow.

ANALOGUE DISPLAY One imitating the movement of a natural timekeeper, such as that of the sun, e.g. an ordinary clock dial with moving hands.

ANCHOR ESCAPEMENT Controller for clocks with pendulums, invented about 1670, which greatly improved timekeeping. In Europe the lever was called an anchor escapement because of its shape.

ARBOR Horological name for an axle, usually it has a toothed wheel fixed to it and both revolve.

ARMILLARY SPHERE Model of Ptolemy's universe with Earth in the centre, made up of bands showing the celestial equator, tropics, ecliptic, etc.

ASTROLABE The most important instrument of the ancient world for astronomy, navigation, and time measurement; it measured the altitude of the stars.

AUTOMATON Animated figure or other object, such as windmill or see-saw.

BALANCE An oscillator – first a bar, then a wheel – for controlling the rate of a timekeeper. Used with a spiral spring after about 1675.

BALANCE COCK The decorative cock holding the bearing of the balance staff at the back of a watch.

BALANCE SPRING A spiral spring used with a circular balance to substitute for gravity. It made the watch much more accurate.

BALANCE STAFF All axles are called arbors except that of the balance, which is called a staff.

BARREL A drum wound with a weighted rope or containing a spring, to provide power.

BEARING HOLE The hole in brass or a jewel, usually ruby, in which a steel pivot turns.

BEAT The number of ticks in a given time. A longcase clock 'beats seconds' i.e. once every second.

BOB The weight at the end of a pendulum rod.

BRIDGE A strap formed like a bridge and fastened at both ends.

BULLE PENDULUM One with a magnetic bob swinging through a coil of wire, invented by M. Favre Bulle in about 1920.

CALENDAR DIAL Special dial giving calendar information such as the day of the week, date, month, year and Church feasts.

CHAFFCUTTER ESCAPEMENT One with two escape wheels, reminiscent of a chaffcutter.

CHAPTER RING The circle bearing the hour numerals or mark (the chapters).

CHECK SPRING One used to check an action, such as that of a bell hammer from resting against the bell.

CHRONOMETER Traditionally a timekeeper with a detent escapement, but used by the Swiss for watches with special timekeeping certificates issued by an official rating bureau.

CLEPSYDRA Greek name for a water clock.

CLICK Horological name for a ratchet pawl, a toothed wheel on which the pawl engages, ensuring motion in one direction.

COCK A part like half a bridge, with only one fixing point.

COMPASS DIAL Sundial depending on the position of the sun in the sky, i.e. the shadow angle, which has to be set north-south.

COMPENSATED Corrected for an error, such as that caused by temperature changes.

CONTRATE WHEEL One with the teeth parallel to the arbor so that it drives through a right angle like a bevel wheel.

COQUERET A steel plate used by French watchmakers instead of an endstone.

COUNT WHEEL A wheel with slots in the edge that controls the number of blows struck in early clocks. Also called a locking plate.

CROSS-BEAT A verge escapement with two bar balances which swing in opposite directions and cross each other, to improve timekeeping.

CROSSING The spoke of a circular balance.

CROWN WHEEL The escape wheel of a verge escapement, like a royal crown, being a band with pointed teeth on one side, parallel to its arbor.

CRUTCH An arm attached to a clock escapement and linking it with the pendulum rod. First used with anchor escapements.

CYLINDER ESCAPEMENT An escapement in which the balance staff is part of a cylinder operated on by the escape wheel teeth.

DART Part of the safety system of a lever escapement, preventing malfunction.

DEAD-BEAT ESCAPEMENT One without recoil, so that if a hand jumps in seconds it does not appear to bounce.

DETACHED ESCAPEMENT One that is not connected with the balance or pendulum except during impulse, e.g. lever and detent escapements.

DETENT A lever that holds up a mechanism until released, i.e. regulates it.

DETENT ESCAPEMENT An accurate dead-beat detached escapement, which employs a detent to hold up the escape wheel.

DIALLING The art of designing the scales of sundials.

DIGITAL DISPLAY One showing the time in numerals, e.g. 07 16 32.

DRAW Geometrical arrangement to prevent a lever escapement from tripping.

DUMB REPEATER A watch that strikes blows on a block inside the case, representing the time, so that the time can be felt. Introduced about 1750 by Julien Le Roy.

DUMBELL BALANCE A bar balance with a knob at each end, used in watches before the balance spring.

DUPLEX ESCAPEMENT One with pairs of teeth on the escape wheel, or two escape wheels. First used for high quality watches, and later low quality ones.

EBAUCHE The plate that forms the frame of the modern watch.

ECLIPTIC The apparent path of the sun around the Earth.

EGLOMISÉ The art of painting a picture on glass for viewing from the opposite side.

ELINVAR A metal which does not change its elasticity at different temperatures, used for temperature compensation of balance springs.

ENDSTONE The gemstone used as an end bearing for a balance staff pivot.

EQUATION OF TIME The difference between solar time (by a sundial) and mean solar time (by a clock).

ESCAPE WHEEL The toothed wheel of a timekeeper by which the escapement controls its rate.

ESCAPEMENT The part of a clock linking the oscillator (balance or pendulum) to the power source (spring or weight) and controlling the rate.

ESSEN RING The first accurate quartz oscillator, used for observatory quartz clocks.

EUREKA BALANCE A very large circular balance used in some early electric clocks.

FAST BEAT ESCAPEMENT One with a fast tick, to improve timekeeping.

FINIAL The decorative terminal of a pillar or other part of a movement or case.

FLAT BED The horizontal frame of a modern turret clock.

FLY The rotating fan used as a governor to control the rate of striking.

FOLIOT An oscillating bar with a weight at each end, used in the first mechanical timekeepers as a controller.

FORM WATCH One with a case in the form of something such as a scarab, violin or cross.

'FRAISE' One of the first milling tools used in watch factories, resembling a strawberry.

FRICTION REST ESCAPEMENT One in which a tooth of the escape wheel rests against a moving part while the balance swings, e.g. the cylinder escapement. It is not detached.

FUSEE The trumpet-shaped pulley used in earlier spring-driven timepieces with a gut line or chain to provide an even power drive.

GNOMON The part of a sundial that throws a shadow.

GOING BARREL A barrel with a spring in it which is wound in the same direction as the drive.

GOING TRAIN The series of gears in the timekeeping side of a clock or watch.

GRANDE SONNERIE Striking the previous hour before striking each quarter.

GRASSHOPPER ESCAPEMENT An ingenious escapement that needed no oiling, devised by John Harrison.

GRAVITY ESCAPEMENT One in which the impulse is supplied by a lever, usually under the influence of gravity – a form of *remontoire*.

GREAT WHEEL The main driving wheel of a clock.

GRIDIRON PENDULUM One made up of a series of rods of two different metals to keep the length constant when the temperature changes.

HAIRSPRING Another name for a spiral balance spring.

HELIO-CHRONOMETER A sundial showing mean time, i.e. the same time as a clock.

HEMICYCLE An early sundial with the scales inside part of a sphere.

HERTZ (HZ) The number of cycles a second of an oscillator. A Royal or seconds pendulum oscillates at 0.5 Hz.

HIPP TOGGLE A 'hit or miss' switch used on some electric pendulum clocks to keep the arc of swing steady.

HIT AND MISS SYNCHRONISER A switch used on a slave pendulum to keep it in step with a free master pendulum.

HORIZONTAL DIAL A fixed compass sundial, e.g. an ordinary garden sundial.

HORIZONTAL ESCAPEMENT Another name for a cylinder escapement.

HORLOGE (horloge, horologium, etc) Latin-derived names for an early timekeeper, hydraulic or mechanical, after the sundial.

HOROLOGY The science of timekeeping.

IMPULSE The tiny push given over a fraction of time to keep an oscillator swinging.

IMPULSE PIN (or pallet) The part by which some oscillators are impulsed.

INCABLOC The most used form of modern shock absorber for watches.

INVAR An iron-nickel alloy which has a negligible coefficient of expansion so does not alter size with temperature changes, used for precision pendulum rods, etc.

ISOCHRONISM Having equal time periods for different arcs of swing.

JACK The earliest form of clock automaton, a human figure striking a bell.

KARRUSEL A more robust form of the tourbillon revolving escapement.

LAND A stretch of metal between other parts.

LANTERN PINION A wheel like a small hamster's wheel used instead of a toothed pinion.

LEAF A tooth of the small driven wheel called a pinion.

LÉPINE CALIBRE A watch movement in which all the wheels are under cocks or bridges.

LEVER ESCAPEMENT The most successful watch escapement, rather like the anchor clock escapement, but detached.

LIGHT-EMITTING DIODE (l.e.d.) A pin-point light source used to make up numerals in electronic digital watches. Not used for continuous display.

LIQUID CRYSTAL DISPLAY (l.c.d.) Figures made up of dark bars on a lighter-coloured ground (or vice versa), used for continuous display in electronic digital watches.

LOCKING PLATE Old name for a count wheel.

MAGNETIC ESCAPEMENT One in which the link between the oscillator and escape wheel is magnetic.

MAINSPRING A coiled spring used as a motive force instead of a weight.

MAINTAINING POWER A means of keeping a timepiece going while being wound.

MASTER AND SLAVE SYSTEM A good timepiece (the master clock) controlling a poorer one, or slave dials showing the time.

MASTER CLOCK See above.

MEAN SOLAR TIME Sundial time corrected by the equation of time so that the hour lengths are equal. A clock shows equal hours, i.e. mean solar time.

MERCURIAL GILDING Covering base metal with powdered gold by mixing it with mercury and burning off the mercury. Superseded by electrogilding.

MERCURIAL PENDULUM One with a jar of mercury for a bob to provide temperature compensation.

MERIDIAN A line of longitude or its projection.

MINUTE WHEEL The toothed wheel, the arbor of which turns the minute hand.

MODULE The operational part of an electronic watch.

MOTION WORK The gearing that drives the hour hand from the minute wheel arbor.

MOVEMENT The 'works' of a mechanical timepiece.

NOCTURNAL A hand-held protractor for telling the time at night by the angle of the Great Bear.

NUTATION A 'nodding' of the Earth as it turns on its axis which has to be corrected for in the time service.

OIGNON A fat watch like an onion, produced in the eighteenth century by French watchmakers.

OIL SINK A countersink around a pivot hole to retain the oil.

ORMSKIRK ESCAPEMENT Another name for the chaffcutter as it was made in Ormskirk, Lancashire.

ORRERY A working model of the movements of the planets, named after Charles, Fourth Earl of Orrery.

OSCILLATOR The controller of a timepiece, e.g. balance, pendulum.

OVERCOIL The outer end of a balance spring curved inwards and over the top of the rest of the spring. to improve timekeeping. Sometimes called a Breguet overcoil. It made the spring 'breathe' evenly on all sides.

PALLET The small shaped part (usually of ruby in watches) through which impulse is imparted to the oscillator.

PARACHUTE The first shock absorber for watches, invented by Breguet.

PEDOMETER WINDING Winding by an eccentric weight that swings with the wearer's movements. Now called self- or automatic-winding.

PENDULE SYMPATHIQUE A clock that will set and regulate a special watch overnight, invented by Breguet.

PENDULUM An oscillator that depends on its length and the force of gravity.

PERPETUELLE The first commercial self-winding watch, made by Breguet.

PILLAR One of the posts that separate the plates of a clock, the whole comprising the frame. Decorative pillars help to date a watch.

PINION The small, steel-toothed wheel driven by a much larger toothed brass wheel.

PIN-LEVER ESCAPEMENT A low-priced lever escapement with steel pin pallets instead of jewelled pallets.

PIN-PALLET ESCAPEMENT An escapement for watches or clocks, with round or D-shaped pallets of steel or ruby acting on the escape wheel.

PIN-WHEEL ESCAPEMENT A French pendulum clock escapement with pins instead of teeth in the escape wheel, used in longcase and turret clock movements.

PIVOT The thin end of an arbor running in a bearing hole.

PLATE Two plates held together by pillars comprise the frame of a clock.

PULL REPEAT A cord in some clocks which, when pulled, will make the clock strike the past hour and perhaps the quarters as well.

PULSE OR PUSH PIECE A projection on a watch case that will 'pulse' to indicate the time silently when required.

QUARTZ OSCILLATOR The timekeeping element of a quartz crystal clock or watch.

RACK A curved or straight arm with teeth.

RACK CLOCK One that falls down a straight rack so that the weight of the clock drives it.

RACK STRIKING Striking controlled by a curved rack and a snail, which cannot get out of phase.

RATCHET WHEEL A toothed wheel on which a pawl engages, ensuring motion in one direction.

RATE The regularity of going of a timekeeper (independent of the time of day).

RATING NUT The nut below a pendulum bob to raise or lower it to alter the rate.

REGULATOR An accurate longcase clock, usually having a special dial with non-concentric hands.

REMONTOIRE A mechanism that loads and reloads a gravity or spring-loaded pallet to provide a constant impulse, instead of applying the impulse direct, to the balance or pendulum.

REPEATER A timekeeper that will sound, or otherwise indicate, the time on demand.

RISE AND FALL A regulator used to adjust timekeeping by raising or lowering the pendulum to alter its length.

ROLLER Part of the safety system to prevent malfunction of a lever escapement.

SAND GLASS Interval timer depending on the flow of sand from one chamber to another.

SCRATCH DIAL Crude vertical sundial scratched on the south porch of a church centuries ago to indicate service times. Sometimes called a Mass dial.

SECONDS PENDULUM One about a metre long beating seconds (after about 1670).

SET-UP Pre-tensioning of a spring barrel used with a fusee.

SOLAR TIME Time from a sundial not corrected by the equation of time.

SLAVE Pendulum, clock, or dial controlled by a master clock.

SNAIL A spiral-shaped cam, involved in the striking movement.

SPIRAL BALANCE SPRING Hairspring used with a circular balance.

SPRING BARREL A driving barrel containing a coiled spring to provide power.

SPRING DETENT Some detents are pivoted; others are mounted on a strip of spring instead.

STACKFREED Very early German friction device to help smooth out the power from a crude mainspring.

STAFF The axle of a balance.

STOP WORK An arrangement to prevent a spring from running down completely.

STRIKING TRAIN The gearing operating the striking work.

STYLE Another name for a gnomon.

SUSPENSION The length of spring, cord, etc. from which a pendulum is supended.

SYNCHRONOUS MOTOR An electric motor that runs in step with the frequency of the electric mains.

TEMPORAL HOURS Division of the period of daylight into so many hours and the period of darkness into the same number of hours. Common before the mechanical clock.

THROW A simple lathe driven by a hand wheel or treadle.

TIME BALL An early time signal indicated by a ball falling down a mast on a tower.

TIME GUN Time signal made by firing a gun.

TIMER A timekeeper indicating the elapse of time instead of time of day.

TIMING SCREWS Screws in the rims of earlier balances for altering the rate.

TING-TANG CHIME One where each quarter is indicated on high and low bells.

TORSION PENDULUM One that twists instead of swinging.

TOURBILLON Rotating escapement invented by Breguet to improve timekeeping.

TRANSDUCER A device transforming electronic oscillations into vibrations or vice versa.

TURRET CLOCK A large public clock, also called a tower clock.

TURNS Early lathe worked by a bow to oscillate the work.

UNIVERSAL RING A sundial adjustable for different latitudes.

VANE An air brake that will slow the movement of a part such as a gravity escapement arm.

VERGE ESCAPEMENT The earliest and most persistent escapement, comprising a staff (the verge) with balance and pallets, and a crown wheel.

VERTICAL DIAL Sundial with a vertical scale.

WAGON SPRING A leaf spring used to power a few American shelf clocks.

WEIGHT BARREL One on which a weighted rope, cable, or gut line is wound to drive a clock.

WHEEL The toothed driving wheel that engages a pinion. Many wheels and pinions are combined on the same arbor.

WINDING CAPSTAN Large medieval church and turret clocks were wound by a spoked capstan or wheel at one end of the barrel.

Bibliography

Abbot, Henry G. *Antique Watches and how to establish their age* (Chicago 1897)
The Watch Factories of America, Past and Present (Chicago 1888)

Allix, C. and P. Bonnert *Carriage Clocks : their history, decoration and mechanism* (Antique Collectors Club 1974)

Andrade, J. *Chronomètrie* (Paris 1908)

Atkins, S. E. and W. H. Overall *Some Account of the Clockmakers' Company* (London 1881)

Bailey, Chris *Two Hundred Years of American Clocks and Watches* (Prentice-Hall Inc., NJ 1975)

Baillie, G. H. *Catalogue of the Museum of the Clockmakers' Company in the Guildhall London* (1949)
Clocks and Watches : an historical bibliography (to 1800) (1951. Repr. Holland and NAG Press, London 1978)
Watches : their history, decoration and mechanism (London 1929. Repr. NAG Press, London 1979)
Watchmakers and Clockmakers of the World (NAG Press, London 1966; USA as Vol 1 1976)

Bain, Alexander *A Short History of the Electric Clock* (1852) Ed. W. D. Hackmann (Turner and Devereux, London 1973)

Basserman-Jordan, Ernst von *The Book of Old Clocks and Watches* 4th ed. rev. Hans von Bertele, trans. Alan Lloyd (George Allen and Unwin, London 1964)

Beckett, E. *A Rudimentary Treatise on Clocks, Watches and Bells for Public Purposes* (1903) (E. P. Group, Yorks 1974)

Beeson, C. F. C. *English Church Clocks 1280–1850* (Antiquarian Horological Society, Kent 1971)

Berner, G. A. *Dictionnaire professionel illustré de l'horlogerie* (Chambre Suisse d'Horlogerie, La Chaux-de-Fonds 1961)
with G. Albert *L'horloger-électricien* (Bienne and Besançon 1926)

Berthoud, F. *Essai sur l'horlogerie* (Paris 1763)
Histoire de la mesure du temps par les horloges (Paris 1802, repr. 1976)

Bird, Anthony *English House Clocks 1600–1850* (David and Charles, Devon 1973; Arco Publishing Co, NY 1973)

Bobinger, Maximilian *Kunstuhrmacher in Alt-Augsburg* (Augsburg 1969)

Britten, F. J. *Old Clocks and Watches and their Makers* 1899, 7th ed. Eds. G. H. Baillie, C. Clutton and C. A. Ilbert 1955 (Eyre Methuen Ltd, London 1973; Charles River Books, Boston, Mass 1976)
Old English Clocks (Wetherfield Collection 1907)
Watch and Clockmaker's Handbook, Dictionary and Guide (1878) 16th ed. Ed. R. Good (Eyre Methuen Ltd, London 1978)

Bromley, John *The Clockmakers' Library* (Sotheby Parke Bernet, London 1977)

Bruton, Eric *Antique Clocks and Clock Collecting* (Hamlyn, Middx 1974)
Clocks and Watches 1400–1900 (Arthur Barker Ltd, London 1967; Praeger Pubs NY 1967)
Clocks and Watches (Hamlyn, Middx 1968)
Dictionary of Clocks and Watches (Arco Publishing Co, NY 1962)
The Longcase Clock (Arco Publishing Co, NY 1964;

Granada Pubs, Herts and NY 1977)

Carrera, R. *Hours of Love* (Scripter SA, Lausanne 1977)

Cescinsky H. and M. Webster *English Domestic Clocks* (London 1913. Repr. Spring Books, Middx.)
Old English Master Clockmakers and their Clocks 1670–1820 (George Routledge, London 1938)

Chamberlain, Paul *It's About Time* (Richard R. Smith, NY 1941)

Chance, B. *Electronic Time Measurement* (Dover Pubs, Essex 1966; Peter Smith, Magnolia Mass.)

Chapuis, A. *Histoire de la Boîte à Musique et de la Musique Mécanique* (Lausanne 1955)
Histoire de la pendulerie Neuchâteloise (Neuchâtel 1917)
La montre automatique ancienne (Neuchâtel 1952)
La montre Chinoise (Neuchâtel 1952)
Urbain Jurgensen et ses continuateurs (Neuchâtel 1924)
with E. Jacquet *Technique and History of the Swiss Watch* (Urs Graf-Verlag, Bern 1953) new edition with new chapter by Samuel Guye (1970)
and E. Droz *Automates* (Neuchâtel 1941; Batsford Ltd, London 1958)
and E. Gelis *Le monde des automates* (Paris 1928)
and E. Jacquet *The History of the Self Winding Watch* (Rolex Watch Co., Geneva 1952)

Cipolla, Carlo M. *Clocks and Culture 1300–1700* (Collins, London 1967; N. W. Norton, NY 1978)

Clark, John E. T. *Musical Boxes : A History and Appreciation* (Allen & Unwin Ltd, London 1961)

Clutton, Cecil Ed. *Collector's Collection* (Antiquarian Horological Society, Kent 1975)
with G. Daniels *Watches* (Batsford Ltd, London 1965)

Coole, P. G. and E. Neumann *The Orpheus Clocks* (Hutchinson Publishing Group, London 1972)

Cumhaill, P. W. *Investing in Clocks and Watches* (Barrie and Rockliff, London 1967)

Cuss, T. P. Camerer *Country Life Book of Watches* (Antique Collectors Club, Woodbridge 1976)

Daniels, George *Art of Breguet* (Sotheby Parke Bernet, London 1974; Biblio Dist. NJ 1975)
English and American Watches (Abelard-Schuman Ltd, London 1967)

Davies, W. O. *Gears for Small Mechanisms* (NAG Press, London 1970)

De Carle, D. *Clocks and Their Value* (NAG Press, London 1968, rev. 1979)
Complicated Watches and their Repair (1956) (Repr. NAG Press, London 1977; Outlet Book Co., NY 1977)
Practical Clock Repairing (1946) (NAG Press, London; Wehman Bros, NJ)
Practical Watch Repairing (1946) (NAG Press, London 1978; Wehman Bros, NJ 1978)
The Watch and Clock Encyclopedia (1950) (Repr. NAG Press, London 1959; Landau Book Co, NY 1976)
Watches and their Value (NAG Press, London 1978)

Defossez, Leopold *Théorie générale de l'horlogerie* (La Chaux-de-Fonds 1950)

Derham, W. *The Artificial Clock Maker* (London 1696)

Ditisheim, Paul, Roger Lallier and L. Reverchon *Pierre le Roy et la Chronométrie* (Tardy, Paris 1940)

Dreppard, Carl W. *American Clocks and Clockmakers* (Charles T. Branford, Mass 1958)

Earnshaw, T. *Longitude : an appeal to the public* (1808)

Ebauches S. A. *Technological Dictionary of Watch Parts* (Neuchâtel 1951)

Eckhardt, George H. *Pennsylvania Clocks and Clockmakers* (Devin-Adair, Conn. 1955)

Edwardes, Ernest L. *The Grandfather Clock* (John Sherratt & Son, Cheshire 1971)
Weight Driven Chamber Clocks of the Middle Ages and Renaissance (John Sherratt & Son, Cheshire 1965)
The Story of the Pendulum Clock (John Sherratt & Son, Cheshire 1977)

Einstein, Albert *Relativity : The Special and General Theory* (Methuen & Co Ltd, London 1920)

Epinasse, Margaret *Robert Hooke* (W. Heinemann Ltd, London 1956)

Ferriday, Peter *Lord Grimthorpe, 1816–1905* (John Murray, London 1957)

Fried, H. B. *The Electric Watch Repair Manual* (Chilton Book Co., Pa 1974)
with Henry B. *The Watch Escapement : the lever, the cylinder, how to analyse, how to repair, how to adjust* (B. Gadow Inc., NY 1974)

Garrard, F. J. *Clock Repairing and Making* (1914)
Watch Repairing, Cleaning and Adjusting (1914)

Gazeley, W. J. *Clock and Watch Escapements* (Butterworth & Co., Kent 1973)
Watch and Clock Making and Repairing (Butterworth & Co., Kent 1970; Transatlantic Arts Inc., NY 1974)

Gelis, Edouard *L'horlogerie ancienne* (Paris 1950)

Georgi, Erasmus *Handbuch der Uhrmacherkunst* (Altona 1867)

Giebel, K. *Das Pendel* (1951)

Glasgow, David *Watch and Clock Making* (1885) (E. P. Group, Yorks 1977; British Book Centre, NY 1977)

Good, R. 'The Accutron' *Horological Journal* (June 1961)
' "Glasses" and glasses' *Horological Journal* (June 1969)
Watches in colour (Blandford Press, Dorset 1978)

Goodrich, W. L. and Hazlitt and Walker *The Modern Clock* (Chicago 1905)

Gordon, G. F. C. *Clockmaking Past and Present* (Technical Press, Oxford 1949)

Gould, R. T. *The Marine Chronometer : its history and development* (1923) (Holland Press, London 1960; Albert Saifer Pubs, NJ)

Gros, Charles *Les échappements d'horloges et de montres* (Paris 1913)

Grossman, J. and H. *Horlogerie théorique* (Bienne 1911–12)

Guye, Samuel and Henri Michel *Time and Space : Measuring Instruments from the 15th to the 19th century* (Pall Mall Press, Oxford 1971)

Haswell, J. Eric *Horology : the science of time measurement and the construction of clocks, watches and chronometers* (London 1947. Repr. E.P. Group, Yorks. 1976; Charles River Books, Boston, Mass 1976)

Hayward, J. F. *English Watches* (Victoria and Albert Museum, London 1956)

Hooke, Robert *Diary* Eds. H. W. Robinson, Walter Adams (Wykeham Pubs., London 1968)

Hope-Jones, F. *Electrical Timekeeping* (NAG Press, London 1940)

Howse, D. and B. Hutchinson *Clocks and Watches of Captain James Cook 1769–1969* (1970)

Huber, Martin *Die Uhren von A. Lange und Sohne Glashuthe/Sachsen* (Munich 1977)

Jagger, Cedric *Paul Philip Barraud* (Antiquarian Horological Society, Kent 1968)

Jendritzki, Hans *The Swiss Watch Repairer's Manual* (Scriptar SA, Lausanne 1964)

Jerome, Chauncey *History of the American Clock Business* (New Haven 1860)

Koestler, Arthur *The Sleepwalkers* (Hutchinson, London 1968; Penguin 1970)

Kurz, O. *European Clocks and Watches in the Near East* (Warburg Institute, London 1975)

Langman, H. R. and A. Ball *Electrical Horology* (1923)

Laycock, W. S. *The Lost Science of John 'Longitude' Harrison* (Brant Wright Assoc., Ashford, Kent 1976)

Lecoultre, F. *A Guide to Complicated Watches* (Bienne 1952)

Lee, R. A. *The First Twelve Years of the English Pendulum Clock, or the Fromanteel Family and their contemporaries* (R. A. Lee, London 1969)
The Knibb Family, Clockmakers (Ernest Benn Ltd, Byfleet 1964)

Leopold, J. H. *The Almanus Manuscript* (Hutchinson &., Co. London 1971)

Lloyd, H. Alan *Chats on Old Clocks* (1951)
The Collector's Dictionary of Clocks (Country Life, Middx 1964)
Some Outstanding Clocks over Seven Hundred Years 1250–1950 (Leonard Hill Books, London 1958)

Loomes, Brian *Watchmakers and Clockmakers of the World* vol 2 (NAG Press, London 1976)
The White Dial Clock (David and Charles, Devon 1974; Drake Pubs. NY 1975)
Country Clocks and their London Origins (David and Charles, Devon 1976; and Vermont 1976)

Marrison, W. A. *The Evolution of the Quartz Crystal Clock* (New York 1948)

Marsh, E. A. *Watches by Automatic Machinery at Waltham* (Chicago 1896)

Maunder E. Walter *The Royal Observatory* (1900)

Maurice, Klaus *Von Uhren und Automaten* (Munich 1968)

Mercer, Vaudrey *John Arnold and Son, chronometer makers* (Antiquarian Horological Society, Ramsgate 1972–5)

Michel, Henri *Scientific Instruments in Art and History* trans. R. E. W. and Francis R. Maddison (1967)

Mody, N. H. N. *Japanese Clocks* (Routledge and Kegan Paul, London 1968; Charles Tuttle, Vermont 1967)

Mortensen, Otto *Jens Olsen's Clock* (Copenhagen 1957)

Mosoriak, Roy *The Curious History of Music Boxes* (Lightner Pub. Co., Chicago 1943)

Mudge, Thomas Jr. *A Description with plates of the Timekeeper invented by the late Mr Thomas Mudge* (1799)

Muhe, Richard and Horand M. Vogel *Alte Uhren* (Munich 1976)

Mussey, Barrows *Young Father Time (Eli Terry)* (Newcomer Society in North America, NY 1950)

Nelthropp, Rev. H. L. *A Treatise on Watch-work* (1873)

Nicholls, A. *Clocks in Colour* (Blandford Press, Dorset 1975; Macmillan, NY 1976)

Nutting, Wallace *The Clock Book* (1924)

Ord-Hume, Arthur W. J. G. *Collecting Musical Boxes and how to repair them* (George Allen & Unwin, London 1967)

Overton, G. L. *Clocks and Watches* (1922)

Palmer, Brooks *Treasury of American Clocks* (Collier Macmillan Pubs., London 1968; Macmillan, NY 1967)

Panicali, Roberto *Watch Dials of the French Revolution 1789–1800* (Scriptar SA, Lausanne 1972)

Pellaton, James C. *Watch Escapements* (NAG Press, London 1949)

Philpott, S. F. *Modern Electric Clocks* (1933)

Planchon, Matthieu *L'évolution du mécanisme de l'horlogerie depuis son origine* (Bourges 1918)

Quill, H. *John Harrison : the man who found longitude* (John Baker Pubs., London 1966)

Rawlings, A. L. *The Science of Clocks and Watches* (Sir Isaac Pitman, NY 1948)

Rees, Abraham *Cyclopedia of Arts and Sciences and Literature* (1819) (David and Charles, Devon 1971; Charles Tuttle, Vermont 1970)

Reid, Thomas *Treatise on Watch and Clockmaking* (Edinburgh 1826)

Robertson, J. Drummond *The Evolution of Clockwork* (E. P. Group, Yorks 1972)

Robinson, T. R. *Modern Clocks, their repair and maintenance* (NAG Press, London 1955)

Roblot, Charles *Collection de cadrans de montres* (Paris 1900)

Royer, Collard F. B. *Skeleton Clocks* (NAG Press, London 1969)

Salomons, Sir David *Breguet* (David Salomons, London 1921)

Saunier, Claudius *Traité des escappements* (Paris 1855. Repr. W. and G. Foyle Ltd, London 1975)
The Watchmaker's Hand Book (Chicago 1894)

Sellink, Dr J. L. *Dutch Antique Domestic Clocks 1670–1870* (H. E. Stenfert, Kroese BV, Leiden 1973)

Shenton, Rita *Christopher Pinchbeck and his family* (Brant Wright Assoc., Ashford, Kent 1976)

Simoni, Antonio *Orologi dal '500 all '800* (Milan 1965)

Smith, Alan *Clocks and Watches* (The Connoisseur, London 1975)

Symonds, R. W. *A History of English Clocks* (King Penguin Books, London 1947)
Thomas Tompion, his life and work (Batsford, London 1951)

Tallis, David *Musical Boxes* (Frederick Muller, London 1971)

Terry, H. *American Clockmaking : its early history and present extent* (1870)

Terwilliger, Charles *The Horolover Collection : 400-day clocks* (NY 1962)

Tripplin, J. *Watch and Clock Making in 1889 Paris Exhibition* (1890)

Tyler, E. J. *Black Forest Clocks* (NAG Press, London 1977)
European Clocks (Ward Lock, London 1969)
The Craft of the Clockmaker (Ward Lock, London 1973; Crown Pubs., NY 1974)

Ullyet, Kenneth *Clocks and Watches* (1971)
In Quest of Clocks (1950. Repr. Spring Books, Middx 1969)
Watch Collecting for Amateurs (Frederick Muller, London 1970)

Ungerer, Alfred *Les Horloges Astronomiques et monumentales* (Strasbourg 1931)

Ward, F. A. B. *Time Measurement : The Science Museum handbook* Pt I *Historical Review*, 4th edition (1958) Pt II *Descriptive Catalogue* 3rd edition (1955)

Willard, John Ware *Simon Willard and His Clocks* (Dover Pubs., Essex 1969; Dover Pubs., Mass 1968)

Williamson, G. C. *Morgan Watch Collection Catalogue*

Wise, S. J. *Electric Clocks* (1948)

Woodcroft, Bennet *Alphabetical Index of Patentees of Inventions (1617–1852)* (London 1854. Repr. Evelyn, Adams and Mackay, London 1969)

Wright, T. D. *Technical Horology* (1906)

Wynter, Harriet and Anthony Turner *Scientific Instruments* (Studio Vista, London 1975; Charles Scribner's, NJ 1976)

HOROLOGICAL SOCIETIES

Antiquarian Horological Society, New House, High Street, Ticehurst, Kent TN5 7AL, England

British Horological Institute, Upton Hall, Newark, Nottinghamshire NG23 5TE, England

National Association of Watch and Clock Collectors, P.O. Box 33, Columbia (Pa) 17512, United States of America

Membership of all of these organizations is available to any one with an interest in horology. Each of them tends to specialize in a different aspect of the subject, but overlaps the main interests of the others. The A.H.S is mostly concerned with European horology and the N.A.W.C.C. with that in the USA. The B.H.I. is active mainly in the technical aspects. In Britain there is also the Worshipful Company of Clockmakers, an ancient guild that still flourishes, but the number of Liverymen is limited by statute. There are several continental societies, some local, the most active of which is probably the Freunde Alter Uhren, 7743 Furtwangen, Ilbenstrasse 54, West Germany.

Index

284

Acknowledgments

The publisher and the author would like to extend their thanks to the organizations, museums and individuals for the pictures which appear on the following pages:

A.D.A.G.P.: 230. Aerofilms Ltd: 237. American Museum, Bath: 143R. Biblioteca Capitolare, Padua: 34. Bibliothèque Nationale, Paris: 16B, 30, 204, 208, 211B. Bodleian Library, Oxford: 7, 13BR, 17T, 23, 47. British Library, London: 38. Trustees of the British Museum, London: 11, 48TL, 57B, 64R, 65R, 68R, 69T, 100T, 100B, 101, 102, 104TL, 104TR, 115, 116TL, 126B, 133B, 134, 144T, 148B, 171B, 225, 226, 242. British Tourist Authority: 262. Eric Bruton: 18T, 21T, 61B, 63, 72TL, 72R, 73L, 73 centre, 86L, 86R, 87B, 89TL, 93BL, 95T, 98T, 104B, 112, 113BL, 113BR, 118, 119B, 121, 124R, 125L, 125R, 133TL, 133B, 136B, 137T, 145B, 146T, 149T, 151L, 153L, 154, 156L, 158, 159BL, 161 centre, 165B, 173L, 173R, 184, 185, 199L, 199R, 200, 202, 222, 228T, 229T, 240T, 255R, 261 centre, 272. Cairo Museum: 13BL. Cambridge University Press: 25L, 25R. Centraal Museum, Utrecht: 106. Christie's: 66, 70, 74BR, 86L, 86R, 87B, 107, 111, 116TR, 117T, 135, 161 centre. Maurice Chuzeville: 57. Henrik Clausen: 258B, 259. Conservatoire National des Arts et Métiers, Paris: 62T, 99T, 99B, 144B. A. C. Cooper: 138, 159T, 159BR. Cooper Bridgeman Library: 157, 231. Salvador Dali Museum, Cleveland, Ohio Collection of Mr and Mrs A. Reynolds Morse: 230. Department of the Environment/Crown Copyright: 28R, 236, 238. Drexel University Museum: 139, 141R. Mary Evans Picture Library: 172, 212. Werner Forman Archive: 12, 27T. Fotomas Index: 60B, 264T, 265T. Clive Friend: 40L, 251. Galerie d'Horlogerie Ancienne, Geneva: 125L, 125R. Gemeente Museum, The Hague: 219B. Germanisches National Museum, Nürnberg: 64. Dr Georg Gerster/John Hillelson Agency Ltd: 10. David Griffiths: 195T, 196. H.M. Postmaster General: 173L. Sonia Halliday Photographs: 16T. Hamlyn Group: 124L, 253T, 253B, 261B. Hirmer Fotoarchiv: 13BL. Historisches Uhren Museum, Wuppertal: 81TL, 82B. Historische Uhrensammlung, Furtwangen: 79L. Michael Holford: 39T, 54T, 67, 87T, 96B, 97R, 143R, 209, 258T, 271. Angelo Hornak: 57T, 122R, 174, 176, 183, 188L, 188R, 189B, 191R, 191L, 197L, 197R, 198, 254. B. Hutchinson: 188L, 188R, 197L, 197R, 198. I.G.D.A.: 19R, 103, 144B, 219B,/G. Dagli Orti: 27B, 62T, 85, 34,/A. Vergani: 37R,/M. Carrieri: 224. Illustrated London News: 152, 170L, 265B, 266B. Ingersoll Ltd: 183, 189B, 191L, 191R. Ivanhoe Studios, Bradford: 42. Cedric Jagger: 136T, 149B. Hubert Josse: 99T, 99B. A. F. Kersting: 37L, 250. Dan Klein Antiques Ltd: 176L. Hans P. Korsgaard: 130B, 137B, 164L. Krajowa Agencja Wydawnicza/Jan Tyminski: 82T, 161T, 162B, 163. Miroslav Krob: 39B, 43T, 108BL, 108BR, 123B. Kulturreportage AB: 245. Kunsthistorischen Museum, Wien: 215. London Museum: 124L. Bill Mason: 247. Mansell Collection: 9, 15, 17B, 147B, 151R, 171T, 177L, 187, 193, 241, 266T. Thomas Mercer Ltd: 166. Metropolitan Museum of Art, New York: 26, 140 (Purchase 1942, Pulitzer Fund), 142L and 143L (Gift of George Coe Graves, 1930). Moravian Gallery, Brno: 123B. Musée Curtius de Liege: 79R. Musée des Beaux Arts, Besançon: 19R, 267, 268L. Museum of Decorative Arts, Prague: 6, 46, 54B, 56R, 58TL, 58TR, 58B, 62 centre, 68T, 76L, 76R, 77L, 78T, 78B, 80, 110L, 110R, 114, 126T. Musée du Louvre, Paris: 57 centre, 65L. Musée de Strasbourg: 240B. Museo del Castello, Milan: 224. Museo del Prado, Madrid: 231. Museo delle Scienze, Firenze: 211T. Museo Nacional de Antropologia, Mexico City: 12. Museum Boerhaave, Leiden: 219T,

256T,/Bibliotheek der Rijksuniversiteit, Leiden: 90T. Museum of Clocks, Sternberk, Moravia: 39B. Museum of Fine Arts, Boston: 142R. Museum of the History of Science, Oxford: 31L, 31R, 32. Muzeum Rzemiosla Artystycznego, Warsaw: 82T, 161T, 162B, 163. Museum of the Worshipful Company of Clockmakers, Guildhall, London: 91B, 94 centre, 97B, 116B, 117B, 119T, 125T, 129T, 131T, 150T, 150B, 170R, 190, 246, 257L. Trustees of the National Gallery, London: 59T. National Maritime Museum, London: 91T, 92R centre, 93LB, 93RT, 94TL, 94TR, 95B, 96TL, 96TR, 97L, 105, 109R, 227T,/Ministry of Defence: 96B. National Monuments Record: 13T, 41R, 264B. National Physical Laboratory/Crown Copyright: 180, 228B. The National Trust: 41L. Naval Museum, Madrid: 85. Newarkes Houses Museum, Leicester: 146T. F. E. Niffle: 79R. Novosti Press Agency: 160. Omega Watch Company: 201R. Osterreichisches Museum für Angewandte Kunst: 81R. Patek Philippe, Geneva: 229B. Private Collection: 51L, 51R, 182, 194, 196, 195T, 195B. Pump Room, Bath: 256B, 257R. Radio Times Hulton Picture Library: 141L, 145T, 168T, 178, 192, 232. T. R. Robinson: 42, 248, 249T, 249B. Ann Ronan Picture Library: 14T, 20BL, 22, 88T, 169, 179L, 206T, 206B, 207B, 213R. Royal Observatory: 97R. Scala: 65L, 211T. Science Museum, London: 24, 29, 53, 54T, 84, 87T, 88B, 89TR, 92T, 103, 113T, 122TL, 167, 168BL, 168BR, 175, 177R, 179R, 181, 209, 213L, 214, 218T, 220R, 227B, 244T, 244B, 268R, 269L, 269R, 270, 271,/Crown Copyright: 8, 14B, 18B, 20BR, 28L, 40R, 50, 69B, 74T, 98B, 108T, 109TL, 210, 217, 218B, 220L, 223, 243. Seiko Time Ltd: 201L, 203. Ronald Sheridan's Photo-Library: 43B, 239. Mike Slingsby: 149B. Smiths Industries Clock Company: 162T. Smithsonian Institution, Washington, D.C.: 35. Snark International: 204, 205, 211B. Sotheby, Parke, Bernet & Co.: 112, 113BL, 123T, 127T, 127B, 129BL, 129BR, 130T, 148T, 164R, 173R, 222. Spink & Son Ltd: 27T. St Edmundsbury Borough Council: 57T, 122R. Staatliche Kunstsammlungen Historisches Museum (Mathematisch-Physikalischer Salon), Dresden: 216. Strike One, London: 83, 138, 159T, 159BR, 166, 174. Studio Meusy, Besançon: 267, 168L. Syndicat d'Initiative de Rouen et de sa Region, Office de Tourisme: 234. Technological Institute, Copenhagen: 273L, 273R. Technical Museum, Košice: 108BL, 108BR. Seth Thomas Co., Talley Industries, Connecticut: 165T. Time Museum, Rockford, Illinois: 153R, 155T, 156R, 186T, 186B, 189T. Tiroler Landesmuseum Ferdinandeum, Innsbruck: 19BL. E. J. Tyler Collection: 157. Gabriel Urbanek: 6, 46, 54B, 56R, 58TL, 58TR, 58B, 62 centre, 68T, 76L, 76R, 77L, 78T, 78B, 80, 110L, 110R, 114, 126T. Victoria and Albert Museum: 67,/Crown Copyright: 60T, 73TR, 74BL. John Wallace: 263T, 263B. Wallace Collection: 77R. John Webb: 136T, 147. Wetherfield Collection: 73L. Woodmansterne Ltd: 40L, 251. Württembergisches Landesmuseum, Stuttgart: 48B, 55, 56L, 75R. Yale University Art Gallery, Connecticut: 153L, 155B (gift of C. Sanford Bull).

The photographs on pages 131LB, 132B, 221, 252, and 260 are reproduced by gracious permission of Her Majesty the Queen.

We are particularly grateful to the staff in the Horological Students Room of the Department of Medieval and Later Antiquities of the British Museum for their kind assistance and co-operation, and to Kenneth D. Roberts of the American Clock and Watch Museum, Bristol, Connecticut. Line drawings by David Penney.